Current Topics in Microbiology and Immunology

135

Editors

Differentiation of B Lymphocytes

Edited by
C.J. Paige and R.H. Gisler

With 25 Figures

Springer-Verlag
Berlin Heidelberg NewYork
London Paris Tokyo

Dr. Christopher J. Paige
Dr. Roland H. Gisler
Basel Institute for Immunology
Grenzacherstrasse 487
CH-4058 Basel

ISBN-13: 978-3-642-71853-3 e-ISBN-13: 978-3-642-71851-9
DOI: 10.107/978-3-642-71851-9

Softcover reprint of the hardcover 1st edition 1987

Library of Congress Catalog Card Number 15-12910

2123/3130-543210

Preface

Rapid progress continues to be made in understanding the molecular and cellular events that comprise B-lymphocyte differentiation. This is due in part to the high level of interest shown by many investigators from diverse disciplines, who find this subject suitable for addressing some of the fundamental issues of immunobiology. B-cell developmental models are being extensively used to investigate cell–cell interactions, molecular mediators of differentiation and proliferation, differential onset of gene programs, and gene rearrangement and expression, as well as the generation of the immune response itself. Not surprisingly, increased understanding of B-cell differentiation sometimes results from the application of new techniques that permit greater insight into the cells comprising the system and the genetic mechanisms by which these cells express their differentiative potential. However, experimental strategies based upon the novel application of established technologies have also led to the clarification of many issues, as well as to the discovery of previously unrecognized problems.

One problem, well recognized by those active in the field, is how to keep up with significant developments as they appear. The purpose of this book, part of a series devoted to analysing current issues in biology, is to help overcome this problem. No attempt at comprehensive coverage of all of the issues has been made. Rather, a more thorough analysis of a few topics is presented. The first section deals with cellular models of differentiation and covers attempts, both in vivo and in vitro, to understand B-cell development. As it has become increasingly apparent that different species may well use different strategies to achieve the requisite degree of antibody diversity needed for the generation of an adequate repertoire, we have included chapters which present the avian and ovine systems in contrast to the standard murine models. The second section deals with molecular analyses of immunoglobulin gene formation, with particular emphasis on variable-region gene utilization. Also discussed are the immunoglobulin class switch during B-cell differentiation and attempts to define genes expressed uniquely during B-cell development.

We thank Leslie Nicklin of the Basel Institute for Immunology for editorial assistance in the preparation of this volume.

Basel, 1987

Christopher J. Paige
Roland H. Gisler

Table of Contents

Indexed in Current Contents

List of Contributors

You will find the addresses at the beginning of the respective contribution

Stromal Cell and Factor-Dependent B Lymphopoiesis in Culture

P.W. KINCADE, P.L. WITTE, and K.S. LANDRETH

1 Introduction

Only a fraction of the cells responsible for humoral immunity (B lymphocytes) are long lived. While some B cells are made by division of other B lymphocytes during the propagation of immune responses, large numbers are continuously produced from immature precursors within hemopoietic tissues such as bone marrow. Recent technological advances have allowed substantial progress to be made in understanding the processes involved in primary B-cell formation and how they are regulated. This is particularly true for the final events in the series because they can be observed in cell culture. On the other hand, findings made with cell cultures provide exceptions to current hypotheses and it is now appropriate to reevaluate those models. Our objectives here are to present recent unpublished findings from in vitro studies, relate these to a general model which has been derived largely from transplantation and whole-animal work, and highlight new avenues of investigation which seem to hold particular promise.

2 A Theoretical Model for B-Lymphocyte Formation

Panels of monoclonal antibodies can be used to identify committed progenitors of B cells, manipulate them experimentally, and construct a probable differentia-

Oklahoma Medical Research Foundation, 825 NE, 13th Street, Oklahoma City, OK 73104, USA

Current Topics in Microbiology and Immunology, Vol. 135

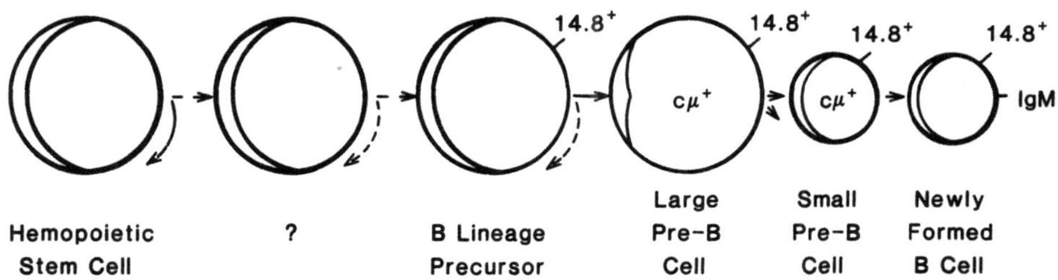

Fig. 1. Possible relationships between precursors of murine B lymphocytes. The model emphasizes the normal expression of an antigen on lymphocytes in adult bone marrow. Normal and leukemic cells in culture provide exceptions to this apparent sequence of events, as detailed in the text

tion sequence for the final steps in de novo lymphopoiesis (KINCADE 1981; KINCADE and PHILLIPS 1985; Fig. 1). Our model was constructed from several experimental findings. All of the precursor cells in adult murine bone marrow which could quickly become functional B cells in culture, and 70% of those which did so in 10-day transfer experiments, could be depleted with our monoclonal 14.8 antibody (KINCADE et al. 1981b; LANDRETH et al. 1983). Most of the lymphocytes in marrow could be highly enriched with this antibody and the suspensions included large and small cells. Approximately one-half of the selected 14.8-bearing (14.8^+) lymphocytes which lacked surface immunoglobulin contained the μ heavy chains of IgM. These "pre-B" cells had earlier been postulated to be the immediate precursors of B cells (RAFF et al. 1976; COOPER 1981). Other studies suggested that large pre-B cells normally divided just once before giving rise to small pre-B cells, which then matured without further division by acquiring surface IgM (LANDRETH et al. 1981). Large cells sorted with a similar monoclonal antibody had extensively rearranged heavy-chain immunoglobulin genes, whereas only the small cells had undergone light-chain gene rearrangement (COFFMAN and WEISSMAN 1983). Not shown in our diagram is the order of acquisition of various other surface markers (Lyb-2, Fc receptors, Th-B, Ia, etc.) which characterize mature, functional B lymphocytes in mice. Variations in assay techniques and the use of different inbred mouse strains make it difficult to integrate the available information in this way.

While such theoretical models have been, and continue to be, most useful for experimental design, some caution is advised about their literal use, especially in interpreting culture results. We have stressed that the initial emergence of B-lymphocyte lineage cells in ontogeny might differ substantially from the process which occurs in adult marrow and have cited examples where initial surface antigen expression was dependent on chronological age, rather than on position in a differentiation sequence (KINCADE 1981, 1984; KINCADE et al. 1982). It was found that most pre-B cells in early embryos lack detectable 14.8 (VELARDI and COOPER, 1984; MEDLOCK et al. 1984), whereas the model indicates that cytoplasmic μ-chains would only be seen in 14.8^+ lymphocytes. After approximately 16 days of gestation, all pre-B cells are 14.8^+ and, while the marker is expressed on a subpopulation of peripheral T cells, there is no evidence that it is associated with myeloid cells (KINCADE et al. 1981b; LANDRETH et al. 1983; SCHEID et al. 1982).

Two other monoclonal antibodies are similar to 14.8, except that they also recognize human lymphocytes. Most pre-B cells in adult human bone marrow were stained by 177.17 antibodies and could be highly enriched with this reagent. However, expression on pre-B cells in neonate marrow was less complete and none of a series of pre-B leukemia cells was found to be positive (LANDRETH et al. 1982; LANDRETH and KINCADE 1982).

Pre-B cells have been studied in regenerating bone marrow and have been found before the appearance of mature B cells (BURROWS et al. 1978; OSMOND and OWEN 1984). An extension of this approach might be expected to provide a suitable test of models such as the one shown in Fig. 1. We recently completed a study of the B-lymphocyte lineage cells in mice recovering from myeloablation caused by 5-fluorouracil, and no apparent order of reappearance was found (V. VETVICKA et al., manuscript in preparation). Persistence of toxic metabolites and their influence on nonlymphoid regulatory cells might well have complicated interpretations in that system. In other unpublished studies we have found that the kinetics of formation of pre-B cells in culture can parallel, rather than precede, expansion of 14.8^+ cells (K.S. LANDRETH, unpublished observations). An even more striking exception to predictions of the model has come from studies of lymphocytes in long-term bone marrow cultures. Pre-B cells grown in these cultures frequently resemble those found during embryogenesis in that they lack 14.8.

Recent studies indicate that a single structural gene encodes the family of cell surface antigens which have been called "leukocyte common antigens," Ly-5, T-200, and B-220 (THOMAS et al. 1985; SHEN et al. 1985). Regulation affecting RNA transcription results in glycoproteins of relative molecular mass (M_r) around 200000 (expressed on most hemopoietic cells and T cells), 205000 (preferentially associated with monocytes), or 220000 (preferentially expressed on B-lineage cells and a subset of peripheral T cells). Many monoclonal antibodies, including 14.8, recognize determinants exclusively associated with the largest molecular form (SCHEID et al. 1982). These presumably are in the N-terminal region of the molecule, and there are probably multiple epitopes (LANDRETH et al. 1982). A similar system has been described in humans, rats, and mice, but little is known of the possible function of these molecules (DALCHAU and FABRE 1981; THOMAS et al. 1985; COFFMAN and WEISSMAN 1981). It will be interesting to learn how such tissue-specific regulation of expression is achieved, but we can conclude that display of the B-cell form (M_r 220000) is not obligate for synthesis of immunoglobulin chains. While pre-B cells do not have to be 14.8^+, virtually all of them in adult marrow are 14.8^+ and it remains to be seen if this correlates with any particular function or ability to mature.

On further investigation of lymphocyte markers, situations may be found where acquisition of one B-lineage antigen is an absolute prerequisite for display of another. It would then be necessary to envision coordinated activation of closely linked genetic loci, or some higher-order mechanism of control of programmed gene expression. In the meantime, the available markers are useful for detecting and manipulating often rare lymphocyte populations and tentatively positioning them in a particular differentiation lineage.

3 General Approaches Involving Lymphoid Cell Cultures

Development of experimental conditions for observing the differentiation of cells in the B lineage has been critical to progress in this field. It may be that the entire process, i.e., from the earliest multipotential stem cells to functional B cells, can only be observed using cell transfer, and several laboratories are making progress in improving clonal markers for such experimental designs (KINCADE and PHILLIPS 1985; KELLER et al. 1985; DICK et al. 1985). However, detailed analyses of the intercellular communication required for the various proliferation and differentiation steps involved can only be made with cell cultures. Before considering our recent progress in this area, it might be useful to review briefly some of the general strategies that have been used.

Perhaps the first B-lineage maturation event to be observed in culture was the division of large bone marrow cells and their transition to small lymphocytes (YOSHIDA and OSMOND 1971; ROSSE 1976). Subsequent investigators exploited very short-term culture systems to investigate the pharmacology of lymphocyte precursor maturation and identify some inducing substances (HAMMERLING et al. 1976; HOFFMAN et al. 1977; SCHEID et al. 1978). However, there were drawbacks to these early studies. First, discontinuous density gradient centrifugation was used for enrichment of the cells of interest, and it was difficult to relate the final results to the incidences of responding cells in the starting populations. In addition, microcytotoxicity was usually used to assess marker expression and this does not necessarily correspond to surface antigen density. Studies done in our laboratory have usually involved changes which can be monitored over several days with a functional assay (cloning ability in mitogen-stimulated, semisolid agar cultures) and/or marker expression (immunofluorescence). We think it important to account for cells through all steps of experimental manipulation and attempt to do this wherever possible. For example, we found that with some hypotonic shock treatments, more than one-half of the nucleated bone marrow cells were lost. Unless such cell loss is shown to be random, the original abundance of the population under study is unclear.

A cloning system developed by METCALF et al. (1975) has been extremely useful in our investigations of B cells and their precursors. A variety of studies have shown that maturation of the immediate precursors of B cells (pre-B cells, cells which contain the heavy chains of IgM) normally occurs inefficiently when the cells are dispersed in semisolid agar cultures (KINCADE et al. 1980). Therefore, functional cells arising in more favorable culture circumstances, e.g., liquid cultures, could be detected by subsequent plating in agar. It was in this way that we obtained our first indication that cellular interactions favor B-lineage precursor maturation in culture (KINCADE et al. 1981a). Semisolid agar cultures also provided an effective system for monitoring emergence of newly formed B cells in transplanted mice of strain CBA/N, which, as a consequence of their genetic defect, never develop clonable cells (KINCADE et al. 1978).

Paige and colleagues extensively modified the agar cloning assay by use of an adherent layer which supports maturation of early precursors (PAIGE et al. 1984, 1985). Because colonies of nonlymphoid cells are abundant in these conditions, a plaquing system was developed to detect terminal maturation of the B cells to a secreting stage (PAIGE and SKARVALL 1982). This provides certain other advantages, such as the ability to investigate light chain selection and allele exclusion at the single-cell level (see SAUTER and PAIGE, this volume). In our own studies, addition of hyperactive regulatory cells or their products enhanced functional maturation of presumed late-stage B-cell precursors so that they were capable of colony formation (see Sect. 9).

Another culture system developed by Klinman supports the transition of pre-B cells to functional B cells (KLINMAN and STONE 1983). B-cell-depleted suspensions are usually transferred to carrier-primed, irradiated recipients. A fraction of the injected cells lodge in the spleen, which is subsequently removed and diced into small fragments. These cells are then cultured with appropriate antigens, and antibody secretion is measured. Advantages of this system, and variations of it, include the ability to do detailed studies of immunoglobulin gene expression by cloned cells. However, there are limitations to the number of precursors which can be studied in a single experiment, and it would be difficult to study cellular and soluble mediator requirements in cultures of irradiated splenic fragments.

4 Long-Term Bone Marrow Cultures

Particularly significant innovations in cell culture methodology were made by WHITLOCK and WITTE (1982; WHITLOCK et al. 1984). They discovered that modification of techniques developed by Dexter for long-term growth of myeloid progenitor cells permitted survival and proliferation of murine B-lymphocyte precursors in strict dependence on cells in an adherent layer. Other investigators have now confirmed their basic observations, although there has been notable laboratory-to-laboratory as well as experiment-to-experiment variation in the types of cells grown (compare for example NAGASAWA et al. 1985; WHITLOCK et al. 1985). Our own recent work has revealed some of the sources of that variability and will be briefly reviewed here. In addition, we hope to focus attention on critical questions which can be addressed with these cultures and their implications for understanding how lymphopoiesis is regulated.

5 Interculture Variability and the Nature of Cultured Lymphocytes

Our experience with long-term cultures began with an appreciation of the extraordinary serum requirements for establishment and maintenance of lymphoid growth. It is common practice to select batches of fetal calf serum that are most suitable for particular types of cell culture, and some variability would be expected with any application. However, we found that successful lymphoid growth was absolutely dependent on stringent serum selection. A total of nine separate batches of fetal calf serum from four different suppliers were tested, and the cultures were evaluated after 3–4 weeks. All sera supported establish-

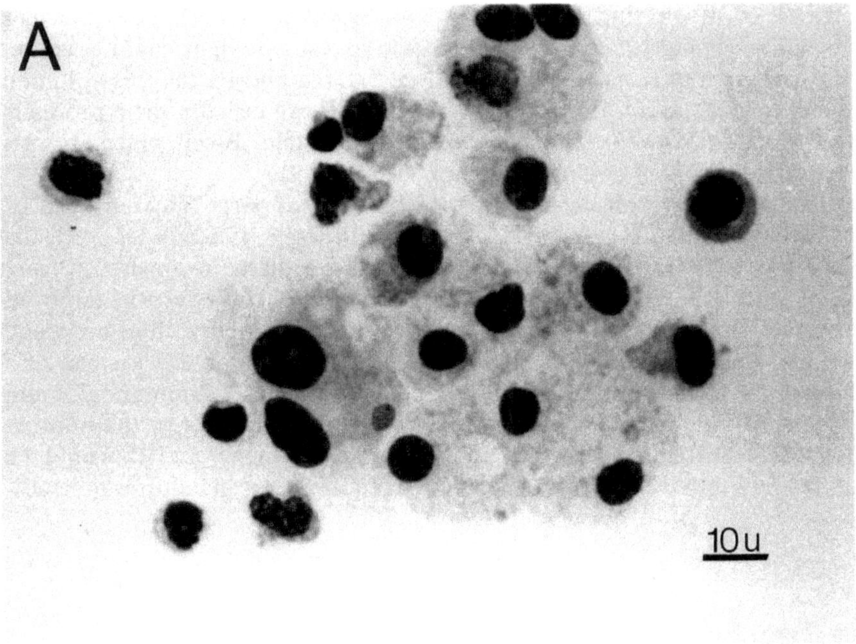
A
10u

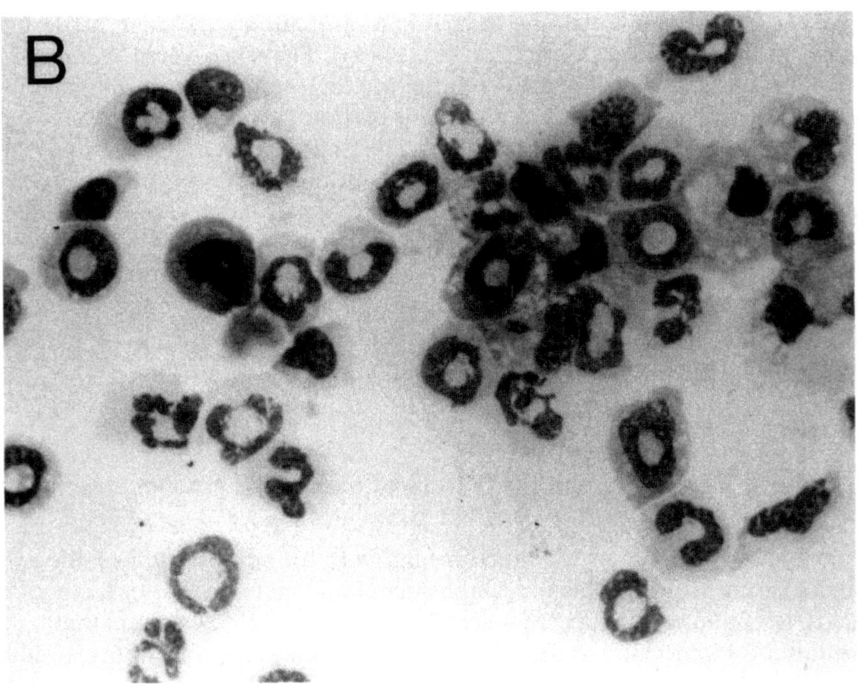
B

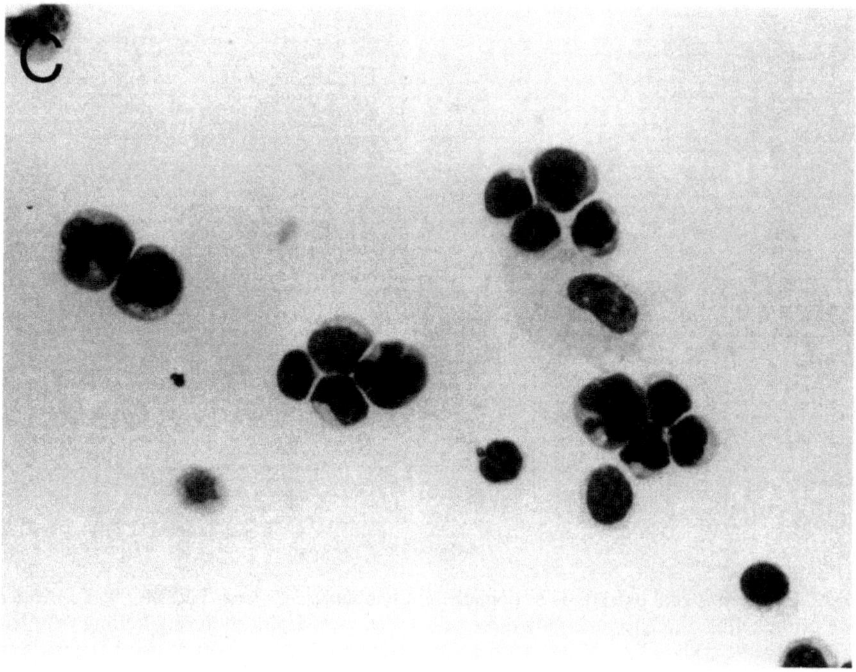

Fig. 2A–C. Outgrowth of nonadherent cells from bone marrow cultured with various batches of fetal calf serum. Cytocentrifuged and Jenner–Giemsa-stained cells are shown from cultures grown in: **A** Irvine Scientific FCS, batch 310642, 8 weeks of culture; **B** Irvine Scientific FCS, batch 311705, 3–4 weeks of culture; **C** Hyclone defined FCS, batch 110451, 3–4 weeks of culture

ment of a nonlymphoid adherent cell layer which appeared pleiomorphic by phase-contrast microscopy. However, the composition of the nonadherent populations differed markedly with the batch of serum used. For example, one batch of serum supported the outgrowth of relatively homogeneous macrophages, others supported predominantly granulocytes or lymphoid cells after 3–4 weeks of culture, and still others sustained multiple cell types (Fig. 2). Thorough characterization of serum components may eventually reveal critical concentrations of hormones or nutrients that are important for establishment of functional adherent layers, as well as for survival and replication of lymphocyte precursors. Until then, this must be empirically determined by exhaustive screening, and variation between batches of serum may constitute a major reason for laboratory-to-laboratory variation.

Proliferating lymphoid cells were seen in our bone marrow cultures only after most myeloid and erythroid cells died and total nonadherent cellularity had decreased about 10-fold. The growth of the lymphoid cells started in small, focal clusters after about 2–3 weeks, and from that time more than 90% of recoverable nonadherent cells had lymphoid morphology. Over the next 1–3 weeks, identifiable B-lineage cells, including 14.8^+ cells, pre-B cells, and B cells, increased in relative frequency and number.

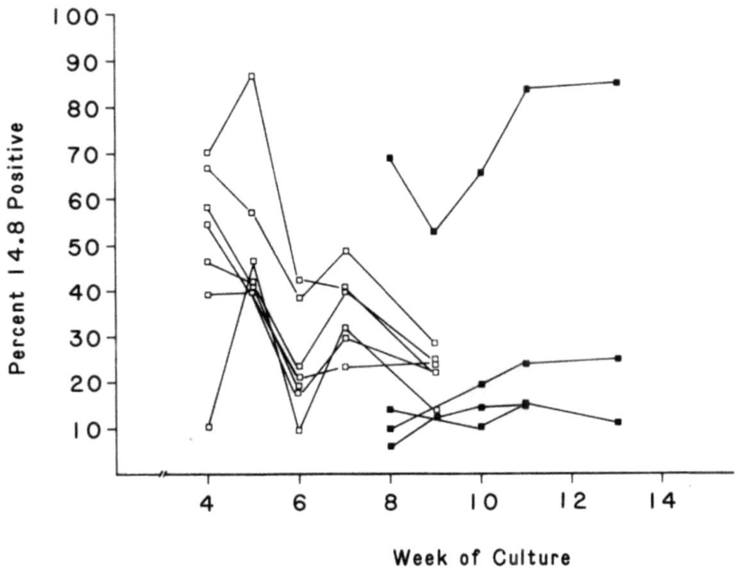

Fig. 3. Changes in 14.8 phenotype of nonadherent lymphoid cells from LTBMC-B. To avoid disruption of the cellular architecture, only those cells which were aspirated during feeding were harvested and analyzed weekly over a 5-week period. Data from replicate cultures of two bone marrow pools are shown

Published reports did not prepare us for the remarkable variability that we found in replicate cultures initiated from a single, pooled suspension of marrow cells. While the kinetics of growth and phenotypes of cells were similar in different culture dishes examined during the first weeks, they had diverged markedly 4–6 weeks after initiation of culture. This divergence was particularly striking for putative early B-lineage cells which express the 14.8 antigen but do not synthesize immunoglobulin chains (Fig. 3; P.L. WITTE et al., to be published). These cells can be from 20% to 80% of the cultured population in replicate cultures. Interestingly, expression of other antigens which might be temporally displayed on developing lymphoid cells has also been variable on our long-term cultured lymphocytes. Some of the markers studied include surface IgD (0%–10%), Ia (0%–10%), ThB (1%–29%), Ly-1 (4%–31%), Mac-1 (1–20%), and Thy-1 (4%–11%).

Serial observations made with many individual culture dishes led to the conclusion that interculture variability became apparent during a discrete period of time (Fig. 3), usually after 4–6 weeks of culture. Subsequently, a more gradual evolution occurred and in later stages (>10 weeks after initiation) the tendency was for incidences of particular lymphocyte sets to stabilize. For example, the composition of one series of replicate lymphocyte cultures from mice of the New Zealand Black (NZB) strain was maintained for one to four cycles of subculturing to new dishes (Fig. 4).

It is interesting to contrast lymphocytes in long-term bone marrow culture (LTBMC) with typical tumor cell lines and mitogen-activated lymphocytes. The

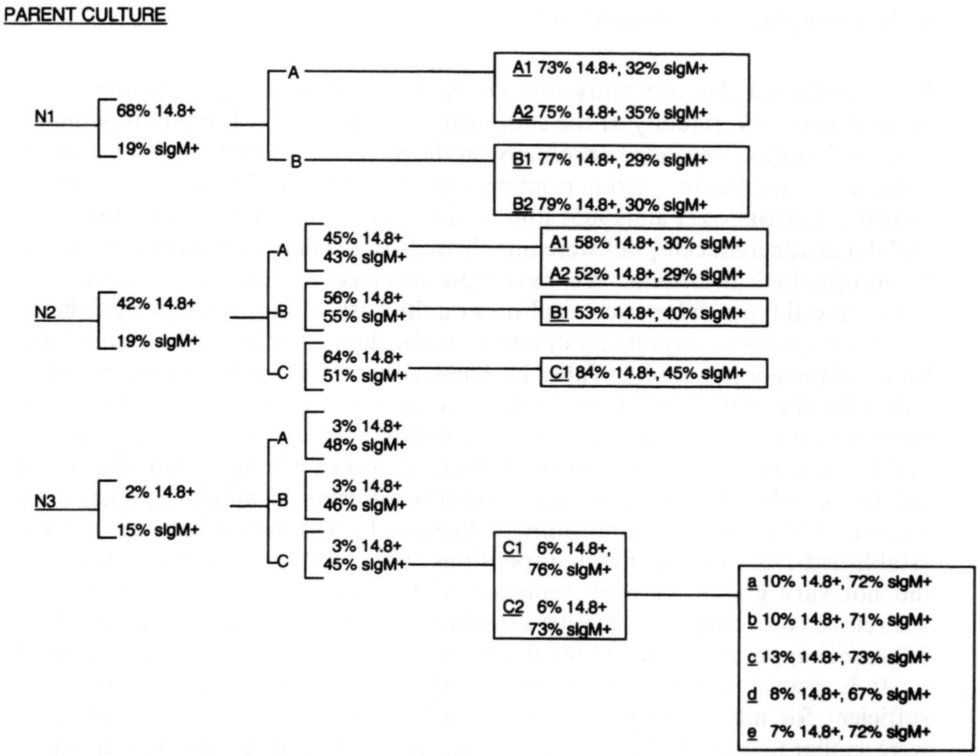

Fig. 4. Stability of B-lineage phenotypes in mature NZB long-term subcultures. Lymphocyte and supporting stromal cells of individual NZB cultures were subcultured into five new cultures, 14 weeks after initiation. Over the next 3 months, cells in the subcultures were repeatedly divided into new replicate cultures. Lymphocyte growth was maintained over three to seven cycles of subculture, but waned after the seventh passage. Periodically, the replicate subcultures were analyzed for their B-lineage composition. Cultures from N1 were analyzed after the third passage, N2 subcultures after the first and third passages and N3 subcultures after the first, third, and fourth passages

latter are blastlike and easily cloned in semisolid agar. Although a comparable fraction of cells in LTBMC are in cycle and mitotic figures can easily be found, they are strikingly similar to small lymphocytes freshly isolated from spleens of normal animals. In addition, the cells maintain a rigid dependence for survival and proliferation on substances produced by, and/or on physical association with, the adherent stromal layer. Different lymphoid compartments can be found in LTBMC and we wondered whether lymphocytes in one compartment might be found to be precursors of those in another. However, careful comparison of lymphocytes which were easily mobilized into the medium (nonadherent) and those which were tightly applied to the adherent layer (stromal-cell-associated) did not reveal significant differences in proliferative activity or relative frequencies of 14.8^+, cytoplasmic-μ-chain-positive ($c\mu^+$), and surface-IgM-positive ($sIgM^+$) cells over 12 weeks of culture.

6 Nonlymphoid Adherent Cells

We investigated the possibility that changes in lymphocyte populations might be mediated by variability of the supporting adherent cells. Long-term lymphocyte proliferation depended absolutely on the presence of adherent bone marrow cells, and supernatants from primary cultures were insufficient to maintain growth. DORSHKIND et al. (1985) studied the intimate association of lymphocytes with large adherent cells, but were unable to distinguish these from macrophages by morphological criteria. We have now resolved at least two predominant adherent cell types in long-term cultures on the basis of morphology, cytochemical staining pattern, expression of receptors for the Fc portion of immunoglobulins, and phagocytic ability (Table 1). These adherent cell subgroups were established by the end of the first week of culture and included macrophages as well as cells characterized by large size, highly spread cytoplasm, oval nuclei, and by their inability to phagocytose small particles or immunoglobulin-coated red blood cells. The latter included subsets differing in alkaline phosphatase content. Although the proportions of these cell types varied between cultures established from different cell suspensions (60%–90% macrophage-like), they did not vary greatly between replicate cultures from the same marrow pool. Therefore, we found no obvious correlation between the composition of the adherent supporting cells and the variation in particular B-lineage cell numbers.

It is crucial to learn which of the adherent cell types are necessary and sufficient for maintaining long-term lymphoid growth. The cells we observed were similar to some of the stromal cells thought to be important for maintaining long-term myeloid cultures (ALLEN and DEXTER 1983, 1984; SONG and QUESENBERRY 1984; ZIPORI et al. 1984; TAVASSOLI and TAKAHASHI 1982). In culture, as well as in situ, very large, alkaline-phosphatase-positive cells have been found

Table 1. Characteristics of adherent cells from long-term culture. Average scores of six cultures of 3 and 8 weeks

	Macrophage-like cells (%)	Large, undefined cells (%)
Esterase positive	95	18
Acid phosphatase positive	90	15
Alkaline phosphatase positive	3	62
Phagocytosis of		
Latex particles	97	0
HEMA particles[a]	69	0
IgG-coated sheep red blood cells	70	0
Pinocytosis of dextran (2 h)	96	0

[a] Fluoresceinated HEMA microspheres (a methacrylate copolymer) were used because of their relatively low nonspecific adherence to cells (FORNUSEK et al. 1983)

in close association with developing myeloid cells (Weston and Bainton 1979). In the lymphoid cultures, similar cells seem to form the foundation for lymphocyte foci. Macrophages surround the periphery of the lymphocyte-laden large cells. Previous studies have indicated the need for macrophage-like cells in the late stages of B-cell differentiation in vitro (Kincade et al. 1981a; Gisler et al. 1984), and this may be the case in long-term culture. However, our preliminary experiments suggest that mature macrophages alone will not support lymphocyte proliferation in long-term cultures.

7 Cells with Potential for Growth in Long-Term Lymphocyte Cultures

The findings summarized above indicate that many stages of the B-lymphocyte lineage are represented in long-term bone marrow cultures. The relative abundance of particular cells is established by unknown processes at a critical stage of culture and can gradually change thereafter. We do not yet know the degree to which differentiation, i.e., the transition from one stage to another, occurs.

Although a number of studies show that lymphocytes from long-term bone marrow culture can reconstitute humoral immunity in vivo (Kurland et al. 1984; Phillips et al. 1984; Nagasawa et al. 1985), questions remain about the nature of the repopulating cells. Kurland et al. (1984) concluded that cultured cells contained bone marrow tropic stem cells, whereas Phillips et al. (1985) found evidence for B-lineage cell reconstitution only in peripheral lymphoid tissues. This indicates that even the relative immaturity of the cells supported in long-term cultures may vary with slightly different conditions.

The typical early pattern of cell decline and subsequent gradual increase in recognizable B-lineage cells suggests that less mature cells establish and give rise to more mature cells. To test this hypothesis, we initiated cultures with bone marrow deficient in B-lineage cells at various stages of differentiation.

Normal bone marrow cells, depleted of mature B cells bearing surface immunoglobulin, or 14.8^+ B-lineage cells, formed pre-B cells and even $sIgM^+$ cells within several weeks of being placed in long-term culture (Table 2). The kinetics of growth and the frequencies of lymphocyte subsets were similar to those seen in cultures initiated with unseparated marrow cells. Fetal liver cells from mice of strain BALB/c of 16 and 17 days gestation also generated $sIgM^+$ B cells in long-term culture when placed on adherent cell layers derived from adult bone marrow (Table 3). In this respect, our data differ somewhat from a previous report (Denis et al. 1984) in that we frequently found mature B cells.

Young NZB mice are notably abnormal in their production of B-lymphocytes (Jyonouchi et al. 1982, 1983; Jyonouchi and Kincade 1984; and see Sect. 9). The composition of lymphoid populations in their marrow is unusual, and there is also reason to suspect dysfunction of microenvironmental regulatory elements. However, in experiments thus far, phenotypic differences in lymphoid cells between long-term cultures established from NZB as compared to BALB/c marrow have not been remarkable. A general paucity of lymphocyte–adherent cell associations (foci) was noted in young adult (and fetal liver cultures) NZB

Table 2. B-lineage cells generated from B-cell-depleted weanling bone marrow tissue in long-term culture

Tissue	Treatment	Culture[b]	Phenotype (%)		
			14.8	Pre-B	B cell
Weanling bone marrow[a]	None	1	54.4	34.0	14.3
		2	41.1	12.1	9.2
		3	43.7	8.8	9.7
		4	3.1	7.3	2.5
		5	60.2	43.9	9.7
	B-cell depleted	1	43.4	47.8	14.0
		2	33.5	21.1	8.9
		3	27.5	21.4	7.9
		4	6.7	24.9	8.8
		5	26.1	47.5	11.5
		6	24.1	28.3	6.0
	14.8 depleted	1	23.9	31.7	10.7
		2	34.0	19.7	14.9
		3	16.5	21.7	7.5
		4	12.5	17.6	9.6
		5	16.8	2.0	7.1
		6	21.1	33.9	12.3

[a] A single pool of bone marrow from 20 mice was prepared and divided into three portions. The cells were incubated on anti-IgM, 14.8, or fetal-calf-serum-coated dishes to remove the appropriate population. The efficiency of removal was determined by fluorescence microscopy (<1% contamination)

[b] Nonadherent lymphocytes from individual cultures were harvested 5–6 weeks after initiation. Cell surface staining was analyzed by a Coulter Epics V cell sorter. Pre-B cells were identified by two-color immunofluorescence (those cells expressing cytoplasmic μ but lacking surface IgM), and scored on a Zeiss photomicroscope III

cultures, and as much as sixfold fewer lymphocytes were found in NZB cultures than in age-matched BALB/c cultures.

Collectively, these studies imply that progenitor cells early in the B lineage can indeed establish in long-term culture. However, the "crisis" stage in newly established cultures is such that a very small number of cells might be responsible for eventual lymphocyte growth. Therefore, such findings do not rule out the possibility that late-stage B-cell precursors may have extensive proliferative potential in vitro. The few residual pre-B cells in the depleted marrow or fetal liver suspensions could have given rise to the pre-B and B cells that were subsequently maintained. The same problem of interpretation would apply to cultures started from cells which are positively selected for in some way.

8 A Cloned Tumor Cell Model

An inducible murine pre-B leukemia line (70Z/3) has been widely used to investigate molecular aspects of immunoglobulin gene expression and factors which

Table 3. B-lineage cells generated from fetal liver in long-term culture

Tissue	Adherent cell source[a]	Culture[b]	Phenotype (%)		
			14.8	Pre-B	B cell
16-day BALB/c fetal liver	BALB/c bone marrow	1	56.0	32.3	12.2
		2	46.0	27.0	22.8
		3	58.0	20.6	6.6
		4	4.0	NT[c]	NT
		5	74.2	27.4	14.9
	NZB bone marrow	1	41.6	18.8	6.7
		2	34.2	7.8	6.0
		3	22.6	32.6	28.6
17-day BALB/c fetal liver	BALB/c bone marrow	1	87.0	10.6	40.4
		2	82.4	14.5	9.7
		3	33.2	18.2	28.5
	NZB bone marrow	1	25.6	NT	NT
		2	6.4	92.3	2.4

[a] Fetal liver cultures were initiated, as reported by DENIS et al. (1984), onto preestablished adherent cell layers from young adult bone marrow. The adherent cell layers were devoid of lymphoid or myeloid cell growth. Nonadherent cells were transferred to fresh monolayers after 7 days of culture
[b] The cultures were harvested 8–9 weeks after initiation. Staining was done as described in Table 2
[c] NT, not tested

influence it. The line was originally described on the basis of its responsiveness to lipopolysaccharide (LPS) by synthesis and display of κ light chains (PAIGE et al. 1978). PAIGE et al. (1982) then demonstrated that it responded in this way to a soluble mediator produced by T cells, and subsequent studies revealed this to be γ-interferon (SIDMAN et al. 1984). γ-Interferon was presumably the inducing agent found in the T-cell-conditioned media employed in several other studies (GIRI et al. 1984; STANTON and MAYNARD 1984; WEEKS et al. 1984). Other natural substances which induce surface IgM expression include interleukin-1 and factors isolated from the serum of young NZB mice (GIRI et al. 1984; STANTON and MAYNARD 1984; JYONOUCHI et al. 1985; and see Sect. 9). The growth of present versions of the cell line is not inhibited by these stimuli, and there is reason to believe that responsiveness only occurs during a portion of the cell cycle (SAKAGUCHI et al. 1980).

ROSOFF and CANTLEY (1983) have investigated some of the biochemical aspects of transmembrane signaling in 70Z/3 cells. The Na^+/H^+ transport inhibitor amiloride inhibited induction of κ by LPS, and some ionophores either enhanced expression or synergized with other stimuli (ROSOFF and CANTLEY 1983). Their clone of the line is responsive to phorbol ester, which causes early changes in free intracellular Ca^{2+} and phosphatidylinositol turnover, and although LPS also stimulated κ synthesis, the early biochemical changes were dissimilar (ROSOFF et al. 1984; ROSOFF and CANTLEY 1985). They concluded that changes in free Ca^{2+} concentration, protein kinase C activation, and transmembrane ion fluxes were all important in determining the responsiveness of the cells.

One of the chromosomes bearing κ light-chain genes is functionally rearranged in 70Z/3 cells, whereas the other allele is in germ-line configuration (MAKI et al. 1980). However, both genes are activated for transcription on exposure of the cells to LPS (PERRY and KELLEY 1979). PARSLOW and GRANNER (1982, 1983) found that an early change occurs in chromatin structure, which exposes nuclease-sensitive sites in the κ-enhancer region of the gene. It is known that in 70Z/3 cells, as well as in other pre-B lines, protein synthesis is not required for LPS-induced activation of κ messenger synthesis (NELSON et al. 1985). This is true even for lines which have both alleles in the unrearranged, germ-line configuration. In fact, brief interruption of protein synthesis can actually induce transcription, and this suggests that κ synthesis may be down-regulated in such cells by a labile repressor protein (WALL et al. 1986).

It is important to note that no stimuli have been shown permanently to induce κ synthesis and sIgM expression in 70Z/3 cells. For example, cloned progeny of LPS- or -interleukin-1-induced cells had only background levels of sIgM when grown out in the absense of stimuli (GIRI et al. 1984). This confirms early observations that surface κ was lost after removal of LPS from bulk cultures of 70Z/3 cells (PAIGE et al. 1978). There are many variants of this tumor, and clones have been deliberately selected which differ in their responsiveness to particular stimuli (MAINS and SIBLEY 1983). This suggests that multiple receptors and/or pathways of induction may be present in pre-B cells, and the variants may prove valuable in dissecting these at a molecular level. There are other examples of murine and human B-lineage tumor cells which are responsive in culture (BOSS et al. 1979; BURCHIEL and WARNER 1980; COSSMAN et al. 1982; LEBIEN et al. 1982). At a minimum, such models provide indicators for soluble mediators which might influence the maturation of normal pre-B cells, and give insight into the sequence of particular differentiation events. For example, 70Z/3 cells provided one of the first indications that $c\mu^+$ pre-B cells must be the immediate precursors of $sIgM^+$ B cells (PAIGE et al. 1978).

9 Soluble Factors Influencing B-Cell Precursors

Having reviewed the major approaches which exploit cell culture methodology, we will now consider the soluble mediators known to be effective in promoting B-cell formation in vitro. Most of the progress in this area has been very recent and our synopsis will not be complete. For example, we will not discuss the effects of interleukin-3 on B-lineage cells, despite impressive claims that this mediator might be sufficient for long-term lymphocyte growth (PALACIOS et al. 1984; PALACIOS and STEINMETZ 1985; MCKEARN et al. 1985). Also conspicuous by their absence will be inhibitors. It seems likely that corticosteroids, prostaglandins, and more specific regulators will counter inducing effects but little information is available on this point. Finally, solubility may not be a necessary requirement for factor activity (KURT-JONES et al. 1985). Exquisite local control would be achieved by direct "presentation" of membrane-bound factors to responding precursors.

Both γ-interferon and interleukin-1 are effective in enhancing maturation of normal murine pre-B cells and 70Z/3 cells in short-term culture (GIRI et al. 1984; SIDMAN et al. 1984; JYONOUCHI et al. 1985). With the selected batches of fetal calf serum that we routinely use, there is a basal level of κ gene expression in the tumor line and, particularly when cells are held in liquid media, some spontaneous maturation of normal murine precursors can be observed. Both γ-interferon and interleukin-1 have effects on multiple organ systems, and interleukin-1 has even been shown to be made by B cells (SCALA et al. 1984). Enormous numbers of B cells (5×10^7) are made daily in mice (OPSTELTEN and OSMOND 1983) and, by extrapolation, in humans (hundreds of billions). It seems unlikely that such systemically acting, lineage-nonspecific mediators are critical regulators of steady-state B-lymphocyte production in vivo. In subsequent studies, we hoped to identify substances which may normally be made in only trace quantities and which could be more restricted in their effects on B-lineage cells.

Examples of genetically determined dysfunction of hemopoietic cells have been particularly valuable for our studies. NZB mice were found to be precocious in terms of maturation of the humoral immune system, and there is hyperactivity associated with B-cell precursors early in life (JYNOUCHI and KINCADE 1984). It appeared from coculture experiments that one or more types of regulatory cells in young NZB bone marrow might be particularly active in promoting the maturation of pre-B cells (JYONOUCHI et al. 1983). Addition of medium conditioned by NZB cells to normal marrow cultures was not revealing. However, the sera of 4-week-old NZB mice had measurable activity. Addition to cultures of as little as a 1:200 final dilution of this material enhanced the functional maturation of precursors from normal marrow. Activity was present only in sera from young NZB mice or irradiated recipients of NZB marrow (JYONOUCHI et al. 1985). Partial purification and further characterization revealed two glycoproteins (JYONOUCHI et al. 1985). One had an isoelectric point (pI) of 3.5 and an apparent M_r of 17000, whereas a slightly smaller (M_r 15000) activity had pIs that varied from 7.8 to 8.4, depending on glycosylation.

Further studies of the biological activity of NZB-serum-derived mediators were done with highly enriched material. No effects on the proliferation of mature B cells, thymocytes, or myeloid cells were found, and indications were that only the final steps in B-cell formation were influenced in culture. While 70Z/3 cells and marrow depleted of B cells bearing surface immunoglobulin were sensitive targets, earlier precursors did not seem to be responsive. That is, when our monoclonal 14.8 antibody was used to remove B cells, pre-B cells, and presumptive earlier cells in the series, no induction of maturation was observed (H. JYONOUCHI unpublished observations).

A quite different type of mediator(s) was discovered from studies of a child with cyclic neutropenia. In addition to the typical fluctuations which occurred in peripheral blood granulocytes, production of pre-B cells in the marrow was periodically dysregulated in this patient (ENGELHARD et al. 1983). At the point in the cycle at which neutrophil numbers were lowest, absolute numbers of pre-B cells were elevated about 50-fold. It was reasoned that a soluble mediator might be associated with this defect and this was sought in samples of the patient's urine (LANDRETH et al. 1985). Concentrated, dialyzed urine induced pre-B-cell production when added to cultures of normal human marrow. This effect was seen only with samples collected on particular days, which corresponded to the predicted increase in the number of pre-B cells in the patient.

Additional unpublished studies revealed that murine hemopoietic cells were responsive to cyclic-neutropenia-associated factor(s), and this permitted a more detailed analysis of the target cells. Immediate precursors of B cells were enriched from normal marrow using monoclonal antibodies coated on plastic dishes (LANDRETH et al. 1983). The survival, proliferation, and maturation of these cells in culture were not obviously influenced by the human material. However, in contrast to all other active substances which have been studied in our laboratory, the cyclic-neutropenia-associated factor(s) induced maturation of early B-lineage precursors. This was indicated by the appearance of $c\mu^+$, 14.8^+ lymphocytes in cultures of marrow which had been extensively depleted with the 14.8 antibody. Optimal induction was observed when marrow from very young animals was cultured for 1–2 days in liquid medium, and the kinetics of emergence of the total 14.8^+ cells tended to parallel increases in pre-B cells. No effects of this kind were noted when NZB serum fractions, interleukin-1, γ-interferon, interleukin-3, or myeloid colony-stimulating activities were tested.

Preliminary characterization of the specimens from the patient with cyclic neutropenia revealed all activity to have a PI of approximately 6. The urinary preparations were fractionated by gel filtration and two peaks of activity were identified (K.S. LANDRETH et al., unpublished observations.)

Activity which eluted with material of M_r 44000–46000 induced pre-B-cell formation in marrow cultures where even adherent cells were depleted. However, a fraction of M_r 15000–17000 was not effective. Therefore, the urinary preparations from patients with cyclic neutropenia contain at least two active components. One may act directly on B-lineage progenitors to effect the generation of pre-B cells, and a second either works via adherent regulatory cells or requires their presence for activity.

10 Synthesis and Concluding Remarks

The first question which arises from our studies is whether a meaningful model can be constructed which describes the steps involved in B-cell formation. Concepts derived from in situ pulse-labeling studies and examination of hemopoietic tissues with monoclonal antibodies led to the scheme that we presented, and this general view remained essentially unchanged for some time (KINCADE 1981). Nonexpression of the epitopes recognized by our monoclonal antibodies on early embryonic pre-B cells, a variety of cultured cells, and transformed human leukemic cells is clearly inconsistent with the adult bone marrow model. These experimental situations may expand normally rare cell sets, and indicate that marrow lymphopoiesis is more complex than we had thought. Alternatively, mechanisms which normally govern display of this highest molecular weight form of common leukocyte antigens could be disrupted by transformation and the conditions of culture. In any event, we have known for some time that it is not obligatory for B-lineage tumor cells to express this antigen, and it will be interesting to learn if it has any normal functional significance. From the standpoint of model building, and for placing unknown cell types in a lineage, it might be advantageous to use multiple cell surface markers and attempt to arrive at a consensus among laboratories.

Results of culture studies provide another apparent exception to the model. Autoradiographic studies indicate that the immediate precursors of B cells and the large cells from which they derive are not extensively selfrenewing (LANDRETH et al. 1983; OPSTELTEN and OSMOND 1983). However, cells with similar characteristics are maintained for prolonged periods in culture. It remains to be determined if these are continually replenished by differentiation of immature progenitors or, alternatively, if they are actively proliferating.

Long- and short-term culture systems offer the most straightforward approaches to learning how B-lymphocyte production is regulated at the molecular level. However, it would be naive to expect the tempo and sequence of events which occur in vitro to be identical to the normal situation in vivo. Moreover, while the cellular composition of cultures can be much simpler than that of fresh bone marrow, certain complexities cannot be minimized. Further technical refinements can be expected and progress should be forthcoming in defining which B-lineage cells can be propagated/induced in culture and which stromal elements are necessary to bring this about. Early studies indicated that components of marrow which adhere to Sephadex G-10 augment B-lineage maturation in culture (KINCADE et al. 1981 a), and we summarized recent findings on their interaction with a cyclic-neutropenia-derived factor. It will be interesting indeed to learn how such poorly defined, but presumably nonlymphoid, elements of hemopoietic stroma relate to adherent cells required in long-term marrow cultures, which are hyperactive in young NZB mice, and which are necessary for the clonable pre-B-cell assay designed by PAIGE et al. (1984, 1985).

We have little information on the cellular origin of the NZB-serum- and cyclic-neutropenia-associated activities. Furthermore, it may prove difficult to obtain these native mediators in sufficient quantity and purity for determination of primary amino acid sequences. However, they represent the first well-documented substances which could be normal regulators of lymphopoiesis, and provide incentive for construction of even better assays for them.

References

Allen TD, Dexter TM (1983) Long term bone marrow cultures: an ultrastructural review. Scan Electron Microsc 4:1851–1866

Boss M, Greaves M, Teich N (1979) Abelson virus-transformed haematopoietic cell lines with pre-B-cell characteristics. Nature 278:551–553

Burchiel SW, Warner NL (1980) Cyclic AMP modulation of Fc receptor expression on a pre-B cell lymphoma. J Immunol 124:1016–1021

Burrows PD, Kearney JF, Lawton AR, Cooper MD (1978) Pre-B cells: Bone marrow persistence in anti-μ suppressed mice, conversion to B lymphocytes, and recovery following destruction by cyclophosphamide. J Immunol 120:1526–1531

Coffman RL, Weissman IL (1981) B220: a B cell specific marker of the T200 glycoprotein family. Nature 289:681–683

Coffman RL, Weissman IL (1983) Immunoglobulin gene rearrangement during pre-B cell differentiation. J Mol Cell Immunol 1:31–38

Cooper MD (1981) Pre-B cells: normal and abnormal development. J Clin Immunol 1:81–89

Cossman J, Neckers LM, Arnold A, Korsmeyer SJ (1982) Induction of differentiation in a case of common acute lymphoblastic leukemia. N Engl J Med 307:1251–1254

Dalchau R, Fabre JW (1981) Identification with a monoclonal antibody of a predominantly B lymphocyte-specific determinant of the human leukocyte common antigen. Evidence for structural and possible functional diversity of the human leukocyte common molecule. J Exp Med 153:753–765

Denis KA, Treiman LJ, St Claire JI, Witte ON (1984) Long-term cultures of murine fetal liver retain very early B lymphoid phenotype. J Exp Med 160:1087–1011

Dick JE, Magli MC, Huszar D, Phillips RA, Bernstein A (1985) Introduction of a selectable gene into primitive stem cells capable of long-term reconstitution of the hemopoietic system of W/W^v mice. Cell 42:71–79

Dorshkind K, Phillips RA (1983) Characterization of early B lymphocyte precursors present in long-term bone marrow cultures. J Immunol 131:2240–2245

Dorshkind K, Schouest L, Fletcher WH (1985) Morphologic analysis of long-term bone marrow cultures that support B-lymphopoiesis or myelopoiesis. Cell Tissue Res 239:375–382

Engelhard D, Landreth KS, Kapoor N, Kincade PW, DeBault LE, Theodore A, Good RA (1983) Cycling of peripheral blood and marrow lymphocytes in cyclic neutropenia. Proc Natl Acad Sci USA 80:5734–5738

Fornusek C, Kopecek J, Vetvicka V (1983) An advantageous method for detection of Fc-receptors and for studying Fc-receptor-mediated phagocytosis. Immunol Lett 7:29–33

Giri JG, Kincade PW, Mizel SB (1984) Interleukin 1-mediated induction of k-light chain synthesis and surface immunoglobulin expression on pre-B cells. J Immunol 132:223–228

Gisler RH, Paige CJ, Hollander G (1984) Role of accessory cells in B lymphocyte progenitor differentiation. Immunobiology 168:441–452

Hammerling U, Chin AF, Abbott J (1976) The ontogeny of murine B-lymphocytes. II. The sequence of B-cell differentiation from Ig^- precursors to plasma cells. Proc Natl Acad Sci USA 73:2008–2012

Hoffman MK, Oettgen HF, Old LJ, Chin AF, Hammerling U (1977) Endotoxin-induced serum factor controlling differentiation of bone-marrow-derived lymphocytes. Proc Natl Acad Sci USA 74:1200–1203

Jyonouchi H, Kincade PW (1984) Precocious and enhanced function maturation of B lineage cells in New Zealand black mice during embryonic development. J Exp Med 159:1277–1282

Jyonouchi H, Kincade PW, Landreth KS, Lee G, Good RA, Gershwin ME (1982) Age-dependent deficiency of B lymphocyte lineage precursors in NZB mice. J Exp Med 155:1665–1678

Jyonouchi H, Kincade PW, Good RA, Gershwin ME (1983) B lymphocyte lineage cells in newborn and very young NZB mice: evidence for regulatory disorders affecting B cell formation. J Immunol 131:2219–2225

Jyonouchi H, Kimmel MD, Lee G, Kincade PW, Good RA (1985) Humoral factors in very young NZB mice that enhance the maturation of normal B cell precursors: partial purification and characterization. J Immunol 135:1891–1899

Keller G, Paige C, Gilboa E, Wagner EF (1985) Expression of a foreign gene in myeloid and lymphoid cells derived from multipotent haematopoietic precursors. Nature 318:149–154

Kincade PW (1981) Formation of B lymphocytes in fetal and adult life. Adv Immunol 31:177–245

Kincade PW (1984) Answerable questions about formation and maintenance of the humoral immune system. Ann Immunol (Paris) 135D:210–215

Kincade PW, Phillips RA (1985) B lymphocyte development. Fed Proc 44:2874–2881

Kincade PW, Moore MAS, Lee G, Paige CJ (1978) Colony forming B cells in F_1 hybrid and transplanted CBA/N mice. Cell Immunol 40:294–302

Kincade PW, Lee G, Scheid MP, Blum MD (1980) Characterization of murine colony-forming

B cells. II. Limits to in vitro maturation, Lyb-2 expression, resolution of IgD$^+$ subsets and further population analysis. J Immunol 124:947–953
Kincade PW, Lee G, Paige CJ, Scheid MP (1981a) Cellular interactions affecting the maturation of murine B lymphocyte precursors in vitro. J Immunol 127:255–260
Kincade PW, Lee G, Watanabe T, Sun L, Scheid MP (1981b) Antigens displayed on murine B lymphocyte precursors. J Immunol 127:2262–2268
Kincade PW, Landreth KS, Lee G (1982) Apparent differences in B-lineage differentiation occurring in fetal and adult life. Ann NY Acad Sci 399:296–303
Klinman NR, Stone MR (1983) Role of variable region gene expression and environmental selection in determining the antiphosphorylcholine B cell repertoire. J Exp Med 158:1948–1961
Kurland JI, Ziegler SF, Witte ON (1984) Long-term cultured B lymphocytes and their precursors reconstitute the B-lymphocyte lineage in vivo. Proc Natl Acad Sci USA 81:7554–7558
Kurt-Jones EA, Beller DI, Mizel SB, Unanue ER (1985) Identification of a membrane-associated interleukin 1 in macrophages. Proc Natl Acad Sci USA 82:1204–1208
Landreth KS, Kincade PW (1982) Differences in the cell surface phenotype of normal and leukemic pre-B cells. In: Vitetta ES (ed) B and T cell tumors: biological and clinical aspects. Academic, New York, pp 57–61
Landreth KS, Rosse C, Clagett J (1981) Myelogenous production and maturation of B lymphocytes in the mouse. J Immunol 127:2027–2034
Landreth KS, Kincade PW, Lee G, Gathings WE, Fu SM (1982) Enrichment of human marrow lymphocytes with monoclonal antibodies to murine antigens. Proc Natl Acad Sci USA 79:2370–2374
Landreth KS, Kincade PW, Lee G, Medlock ES (1983) Phenotypic and functional characterization of murine B lymphocyte precursors isolated from fetal and adult tissues. J Immunol 131:572–580
Landreth KS, Engelhard D, Beare MH, Kincade PW, Kapoor N, Good RA (1985) Regulation of human B lymphopoiesis: effect of a urinary activity associated with cyclic neutropenia. J Immunol 134:2305–2309
LeBien TW, Bollum FJ, Yasmineh WG, Kersey JH (1982) Phorbol ester-induced differentiation of a non-T, non-B leukemic cell line: model for human lymphoid progenitor cell development. J Immunol 128:1316–1320
Mains PE, Sibley CH (1983) LPS-nonresponsive variants of mouse B cell lymphoma, 7OZ/3: isolation and characterization. Somatic Cell Genet 9:699–720
Maki R, Kearney J, Paige C, Tonegawa S (1980) Immunoglobulin gene rearrangement in immature B cells. Science 209:1366–1369
McKearn JP, McCubrey J, Fagg B (1985) Enrichment of hematopoietic precursor cells and cloning of multipotential B-lymphocyte precursors. Proc Natl Acad Sci USA 82:7414–7418
Medlock ES, Landreth KS, Kincade PW (1984) Putative B lymphocyte precursor cells in early murine embryos. J Dev Comp Immunol 8:887–894
Metcalf D, Nossal GJV, Warner NL, Miller JFAP, Mandel TE, Layton JE, Gutman GA (1975) Growth of B-lymphocyte colonies in vitro. J Exp Med 142:1534–1549
Nagasawa R, Kanagawa O, Tittle TV, Chiller JM (1985) In vivo maturation of pre-B cells derived from long-term cultured bone marrow. J Immunol 135:965–972
Nelson KJ, Kelley DE, Perry RP (1985) Inducible transcription of the unrearranged k constant region locus is a common feature of pre-B cells and does not require DNA or protein synthesis. Proc Natl Acad Sci USA 82:5305–5309
Nishikawa S-I, Kina T, Amagai T, Katsura Y, (1985) Early B cell precursors in long-term bone marrow culture selective development in the bone marrow or irradiated recipients. Eur J Immunol 15:696–701
Opstelten D, Osmond DG (1983) Pre-B cells in mouse bone marrow: Immunofluorescence stathmokinetic studies of the proliferation of cytoplasmic μ-chain bearing cells in normal mice. J Immunol 131:2635–2640
Osmond DG, Owen JJT (1984) Pre-B cells in bone marrow: size distribution profile, proliferative capacity and peanut agglutinin binding of cytoplasmic μ chain-bearing cell populations in normal and regenerating bone marrow. Immunology 51:333–342
Paige CJ, Skarvall H (1982) Plaque formation by B cell colonies. J Immunol Methods 52:51–61
Paige CJ, Kincade PW, Ralph P (1978) Murine B cell leukemia line with inducible surface immunoglobulin expression. J Immunol 121:641–647

Paige CJ, Schreier MH, Sidman CL (1982) Mediators from cloned T helper cell lines affect immunoglobulin expression by B cells. Proc Natl Acad Sci USA 79:4756–4760

Paige CJ, Gisler RH, McKearn JP, Iscove NN (1984) Differentiation of murine B cell precursors in agar culture. Frequency, surface marker analysis and requirements for growth of clonable pre-B cells. Eur J Immunol 14:979–987

Paige CJ, Skarvall H, Sauter H (1985) Differentiation of murine B cell precursors in agar culture. II. Response of precursor-enriched populations to growth stimuli and demonstration that the clonable pre-B cell assay is limiting for the B cell precursor. J Immunol 134:3699–3704

Palacios R, Steinmetz M (1985) IL3-dependent mouse clones that express B-220 surface antigen, contain IG genes in germ-line configuration, and generate B lymphocytes in vivo. Cell 41:727–734

Palacios R, Henson G, Steinmetz M, McKearn JP (1984) Interleukin-3 supports growth of mouse pre-B cell clones in vitro. Nature 309:126–131

Parslow TG, Granner DK (1982) Chromatin changes accompany immunoglobulin k gene activation: a potential control region within the gene. Nature 299:449–451

Parslow TG, Granner DK (1983) Structure of a nuclease-sensitive region inside the immunoglobin kappa gene: evidence for a role in gene regulation. Nucleic Acids Res 11:4775–4592

Perry PP, Kelley DE (1979) Immunoglobulin messenger RNAs in murine cell lines that have characteristics of immature B lymphocytes. Cell 18:1333–1339

Phillips RA, Bosma M, Dorshkind K (1984) Reconstitution of immune-deficient mice with cells from long-term bone marrow cultures. In: Wright DG, Greenberger JS (eds) Long-term bone marrow culture. Liss, New York, p 309

Phillips RA, Bosma M, Dorshkind K (1985) Reconstitution of immune-deficient mice with cells from long-term bone marrow cultures. KROC Found Ser 1984; 18:309–321

Raff MC, Megson M, Owen JJ, Cooper MD (1976) Early production of intracellular IgM by B-lymphocyte precursors in mouse. Nature 259:224–226

Rosoff PM, Cantley LC (1983) Increasing the intracellular Na^+ concentration induces differentiation in a pre-B lymphocyte cell line. Proc Natl Acad Sci USA 80:7547–7550

Rosoff PM, Cantley LC (1985) Lipopolysaccharide and phorbol esters induce differentiation but have opposite effects on phosphatidylinositol turnover and Ca^{2+} mobilization in 7OZ/3 pre-B lymphocytes. J Biol Chem 260:9209–9215

Rosoff PM, Stein LF, Cantley LC (1984) Phorbol esters induce differentiation in a pre-B lymphocyte cell line by enhancing Na^+/H^+ exchange. J Biol Chem 259:7056–7060

Rosse C (1976) Small lymphocyte and transitional cell populations of the bone marrow; their role in the mediation of immune and hemopoietic progenitor cell functions. Int Rev Cytol 45:155–290

Sakaguchi N, Kishimoto T, Kikutani H, Watanabe T, Yoshida N, Shimizu A, Yamawaki-Kataoka Y, Honjo T, Yamamura Y (1980) Induction and regulation of immunoglobulin expression in a murine pre-B cell line, 7OZ/3. 1. Cell cycle-associated induction of sIgM expression and k-chain synthesis in 7OZ/3 cells by LPS stimulation. J Immunol 125:2654–2659

Scala G, Kuang YD, Hall RE, Muchmore AV, Oppenheim JJ (1984) Accessory cell function of human B cells. I. Production of both interleukin 1-like activity and an interleukin 1 inhibitory factor by an EBV-transformed human B cell line. J Exp Med 159:1637–1652

Scheid MP, Goldstein G, Boyse EA (1978) The generation and regulation of lymphocyte populations. Evidence from differentiative induction systems in vitro. J Exp Med 147:1727–1743

Scheid MP, Landreth KS, Tung JS, Kincade PW (1982) Preferential but nonexclusive expression of macromolecular antigens on B-lineage cells. Immunol Rev 69:141–159

Shen F-W, Saga Y, Litman G, Freeman G, Tung J-S, Cantor H, Boyse EA (1985) Cloning of Ly-5 cDNA. Proc Natl Acad Sci USA 82:7360–7363

Sidman CL, Paige CJ, Schreier MH (1984) B cell maturation factor (BMF): a lymphokine or family of lymphokines promoting the maturation of B lymphocytes. J Immunol 132:209–222

Stanton T, Maynard M (1984) Cytokine effects on a B-Cell line. Immunol Lett 7:301–307

Song GX, Quesenberry PJ (1984) Radioresistant murine marrow stromal cells: a morphologic and functional characterization. Exp Hematol 12:523–533

Tavassoli M, Takahashi K (1982) Morphological studies on long-term culture of marrow cells: characterization of the adherent stromal cells and their interactions in maintaining the proliferation of hemopoietic stem cells. Am J Anat 164:91–111

Thomas ML, Barclay AN, Gagnon J, Williams AF (1985) Evidence from cDNA clones that the rat leukocyte-common antigen (T200) spans the lipid bilayer and contains a cytoplasmic domain of 80,000 Mr. Cell 41:83–93

Velardi A, Cooper MD (1984) An immunofluorescence analysis of the ontogeny of myeloid, T and B lineage cells in mouse hemopoietic tissues. J Immunol 133:672–677
Wall R, Briskin M, Carter C, Govan H, Taylor A, Kincade P (1986) Labile inhibitor blocks gene transcription in pre-B leukemic cell line. Proc Natl Acad Sci USA 83:295–298
Weeks RS, Mains PE, Sibley CH (1984) Comparison of membrane IgM expression in the murine B cell lymphoma 7OZ/3 treated with LPS or supernatant containing T cell factors. J Immunol 133:351–358
Westen H, Bainton DF (1979) Association of alkaline-phosphatase-positive reticulum cells in bone marrow with granulocytic precursors. J Exp Med 150:919–937
Whitlock CA, Witte ON (1982) Long-term culture of B lymphocytes and their precursors from murine bone marrow. Proc Natl Acad Sci USA 79:3608–3612
Whitlock CA, Robertson D, Witte ON (1984) Murine B cell lymphopoiesis in long-term culture. J Immunol Methods 67:353–369
Whitlock CA, Denis K, Robertson D, Witte ON (1985) In vitro analysis of murine B-cell development. Ann Rev Immunol 3:213–235
Yoshida Y, Osmond DG (1971) Identity and proliferation of small lymphocyte precursors in cultures of lymphocyte-rich fractions of guinea pig bone marrow. Blood 37:73–86
Zipori D, Friedman A, Tamir M, Silverberg D, Malik Z (1984) Cultured mouse marrow cell lines: Interactions between fibroblastoid cells and monocytes. J Cell Physiol 118:143–152

Long-Term Murine Hemopoietic Cultures as Model Systems for Analysis of B Lymphocyte Differentiation

K. DORSHKIND[1] and O.N. WITTE[2]

1 Introduction

The development of B lymphocytes in the bone marrow proceeds in parallel with that of other hemopoietic lineages (ROSSE 1976). Phenotypic analysis using antibodies that detect immunoglobulin (Ig; OSMOND and NOSSAL 1974; ABNEY et al. 1978; LANDRETH et al. 1981; RAHAL and OSMOND 1984) and other nonIg cellular determinants (reviewed by SCHEID et al. 1982) has identified cells in the B-lymphocyte lineage and contributed greatly to the delineation of stages of that process. However, these studies have not always identified the earliest stages of B lymphopoiesis, or permitted regulatory events that occur during B-cell differentiation to be dissected. In order to facilitate such investigations, considerable efforts have been made to develop in vitro systems that permit all phases of primary B-cell development to be studied under controlled conditions.

Desirable objectives in the development of such systems are to focus on B lymphopoiesis from its earliest stages and to permit its regulation to be studied at both the molecular and cellular levels. Initial success at attaining B-cell differentiation in vitro was reported using fetal liver (OWEN et al. 1974; MELCHERS

[1] Division of Biomedical Sciences, University of California, Riverside, CA 92521-0121, USA
[2] Department of Microbiology and Molecular Biology Institute, University of California, Los Angeles, CA 90024, USA

Current Topics in Microbiology and Immunology, Vol. 135

1977), and the recently described double-agar system of Paige appears to detect a very early precursor cell in that tissue (PAIGE 1983; PAIGE et al. 1984). It has also been possible to obtain in vitro differentiation of B lymphocytes from precursors in adult bone marrow (LAU et al. 1979; KINCADE et al. 1981; LANDRETH et al. 1983). These latter cultures have permitted various cellular regulatory events that occur during B-cell differentiation to be investigated. Another relatively recent innovation whose potential for analysis of primary B-cell production is only now being realized is the long-term culture system. The key feature of long-term cultures is the establishment in vitro of an adherent layer of cells, thought to duplicate the in vivo hemopoietic microenvironment that supports long-term hemopoiesis for periods of from several months to a year. Different long-term cultures have been established using bone marrow (DEXTER et al. 1977a; WHITLOCK and WITTE 1982) or fetal liver (DENIS et al. 1984), and all are dynamic systems in which the continuous growth and differentiation of hemopoietic cells occurs. Generally, these cultures minimize the considerable complexity of fresh bone marrow and have allowed both cellular and molecular techniques to be applied to the analysis of B-cell differentiation in vitro.

This chapter will be restricted to describing the different long-term culture systems and will specifically focus on their relevance to B lymphopoiesis. The various short-term in vitro systems that permit the production of B cells from precursors in fetal liver (OWEN et al. 1974; MELCHERS 1977) or bone marrow (LAU et al. 1979; KINCADE et al. 1981; LANDRETH et al. 1981) will not be discussed.

2 Hemopoietic Lineages Present in Long-Term Cultures

Two principle types of long-term bone marrow culture have been described. The system defined by DEXTER et al. (1977a) maintains myeloid and lymphoid precursors, but the only differentiated cells in these myeloid bone marrow cultures (MBMC) are granulocytes and macrophages. Myelopoiesis does not occur in the cultures described by WHITLOCK and WITTE (1982); the predominant cells in these lymphoid bone marrow cultures (LBMC) are mononuclear cells. A third type of long-term culture is initiated with fetal liver (FLC; DENIS et al. 1984), and is dependent for growth on a bone-marrow-derived adherent layer. This section will review the hemopoietic lineages present in these cultures (Table 1); the following section will describe the supporting cells in the adherent layer.

2.1 Myeloid Bone Marrow Cultures

Myeloid bone marrow cultures are initiated by culturing the contents of a femur in medium supplemented with horse serum and corticosteroids, and by 2 weeks after initiation, a confluent, adherent layer has established in the culture flask. While this layer is being formed, those hemopoietic cells dependent on it for

Table 1. Characteristics of long-term hemopoietic cultures

Culture	Culture conditions	Source of adherent layer	Source of hemopoietic cells	Types of hemopoietic cells present
MBMC[a]	Medium (e.g., α-minimal essential medium), 20% horse serum, 10^{-7} *M* steroids, antibiotics, 33°C	Bone marrow	10^6–10^7 fresh marrow cells from 4–12-week-old mice used as recharge after adherent layer established	CFU-S, CFU-GM, all myeloid progenitors; B- and T-cell precursors, no mature B or T cells
LBMC[b]	RPMI-1640, 5% fetal calf serum, 37°C	Bone marrow	3–4-week-old bone marrow from initial inoculum; no recharge necessary	B cells, pre-B cells, lymphoid progenitor?
FLC[c]	RPMI-1640, 5% fetal calf serum, 37°C	Bone marrow	14–17-day fetal liver plated onto preestablished feeder layer	Progenitor B cells

[a] Myeloid bone marrow cultures of DEXTER
[b] Lymphoid bone marrow cultures of WHITLOCK and WITTE (1982)
[c] Fetal liver cultures of DENIS and WITTE (1986)

growth often die. Therefore, it is customary to recharge the cultures with fresh bone marrow cells. The age of the cultures is described in weeks after recharge, and sustained hemopoiesis can be observed for several months. There are variations in the initiation of these cultures. For example, fetal calf serum has been used instead of horse serum provided that exogenous steroids are present, and recharge of the cultures is not always essential.

Hemopoietic cells in MBMC can be demonstrated in a population of nonadherent cells that float in the supernatant medium upon agitation of cultures, and that are attached to and interspersed among the adherent supporting cells. The cells from the adherent fraction grow in large clusters containing several hundred cells in close apposition (Fig. 1a). Up to 95% of the hemopoietic cells in either fraction are, from morphological criteria, neutrophilic granulocytes (Fig. 1b). This is consistent with the observation that granulocyte–macrophage progenitors (CFU-GM), spleen colony-forming units (CFU-S), and mixed-colony-forming cells (CFU-GEMM) are also present in the cultures (KELLER et al. 1983; DEXTER et al. 1977c). Progenitor cells for all other myeloid lineages can also be detected (KELLER et al. 1983; DEXTER et al. 1977c; WILLIAMS et al. 1978).

The feature that most distinguishes MBMCs from fresh marrow is the absence, by morphologic or functional criteria, of lymphocytes. The lack of T cells is indicated by the inability of cultured cells to cause graft-versus-host disease when injected into irradiated allogeneic hosts (DEXTER et al. 1977b; DEXTER and SPOONCER 1978), although manipulation of culture conditions does permit survival of marrow T cells (DORSHKIND and PHILLIPS 1982), and possibly their progenitors (SCHRADER et al. 1979). Cells from the cultures do not respond

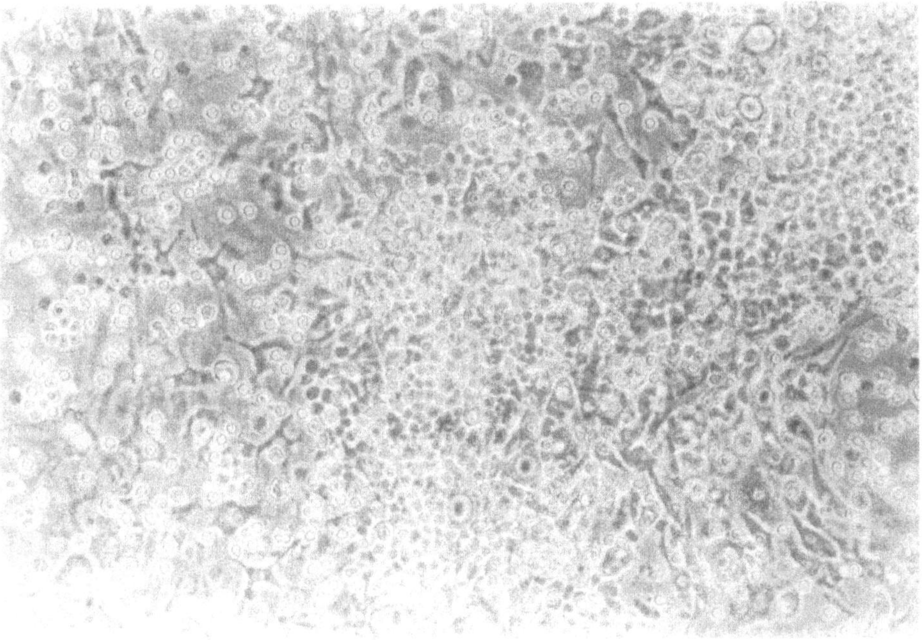

a

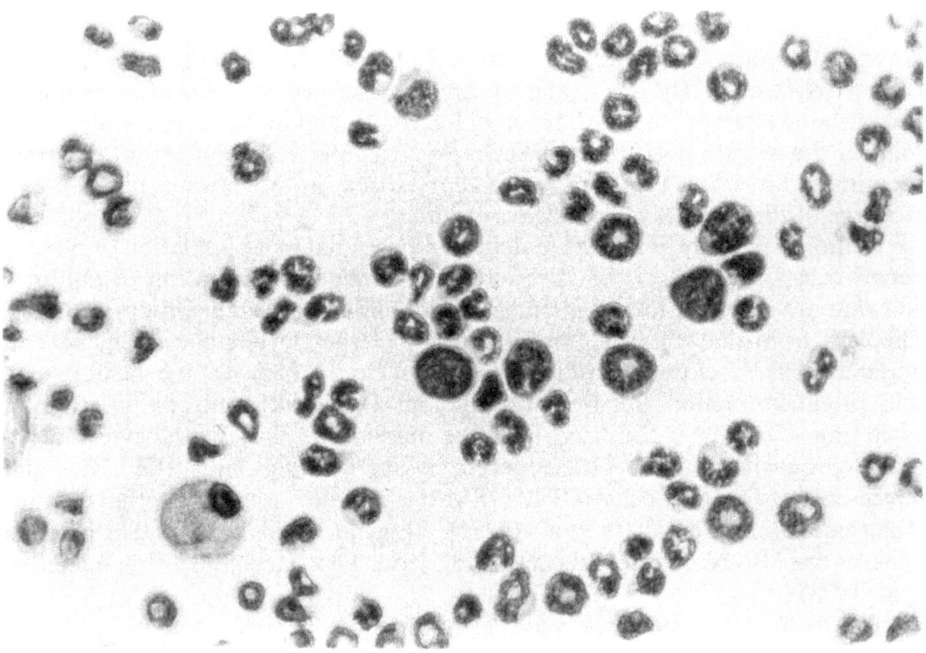

b

Fig. 1. a Phase-contrast micrograph of myeloid bone marrow culture. A large, densely packed cell aggregation is shown with a few interspersed fat cells. **b** Nonadherent cells from myeloid bone marrow cultures. Most cells are granulocytes, although two blast cells are present. (From DORSHKIND et al. 1985)

to B-cell mitogens (K. Dorshkind, unpublished observation) and do not generate B-lymphocyte colonies (CFU-B; Phillips et al. 1984; Dorshkind 1986). Furthermore, no cells that express the 14.8 pre-B-cell antigen or cytoplasmic Ig have been detected (Dexter and Spooncer 1978; Dorshkind 1986). Most of the above reports are in agreement about the absence of cells expressing surface Ig in MBMC, although another study (Aspinall and Owen 1983) has indicated that up to 0.5% of cells express surface Ig and survive in the cultures for up to 6 weeks. These latter cultures were initiated under bulk conditions, and the differences in size, volumes, and plating conditions could account for the observation.

All studies agree that B- and T-cell precursors are present in the cultures. Schrader and Schrader (1978) reconstituted irradiated F_1 hybrid mice with cells from parental bone marrow cultures and used alloantigenic differences to confirm donor engraftment. The reconstituted B cells were fully functional and could respond to a panel of antigens. Reconstitution of B cells by cells from F_1 cultures transplanted into irradiated parental mice was also demonstrated by Aspinall and Owen (1983). Jones-Villeneuve and Phillips (1980) used cells from MBMC initiated with chromosomally marked T6 marrow to reconstitute B cells in irradiated mice, and observed that recipient lymphocytes stimulated to blastogenesis by B- and T-cell mitogens bore that marker. Two interesting observations arose from that experiment. First, in some recipient mice, T6-marked cells were noted in lymphoid but not myeloid cells. This suggested that distinct lymphoid and myeloid precursors were present in the cultures and that selective reconstitution occurred. Secondly, a close association existed between B- and T cell repopulating ability, since the proportion of marked cells in blasts sensitive to lipopolysaccharide (LPS) or concanavalin A (ConA) was always comparable. This was interpreted as further evidence for the existence of a lymphoid-restricted stem cell, rather than distinct B- and T-cell progenitors, in the cultures.

A more extensive characterization of the B-cell precursor in MBMC was performed by Dorshkind and Phillips (1983). This study used immunodeficient CBA/N mice as recipients of cells from cultures established with normal, syngeneic CBA/J bone marrow. CBA/N mice have an X-linked immune defect, one result of which is an inability of cells from their hemopoietic tissues to grow CFU-B. The microenvironment in these mice is normal, since CFU-B can be reconstituted by grafts of normal B-cell precursors (Kincade 1977). By measuring the levels of CFU-B in irradiated, MBMC-reconstituted CBA/N mice, it was demonstrated that cells of the adherent fraction increased CFU-B more than the nonadherent fraction did. One explanation for this was that cells of the adherent fraction have a greater proliferative potential than do ones from the nonadherent fraction, a finding consistent with observations for CFU-S in the two populations (Mauch et al. 1980). Another possibility is that there are more B-cell precursors in the adherent fraction. This possibility must be considered, since a linear relationship exists between the number of precursors transplanted and the amount of B-cell reconstitution observed. A tantalizing possibility from these characterization studies is that the B-cell precursor in the cultures may be more primitive than the pre-B cells previously studied.

This idea is based on observations that the cultured cells had delayed kinetics of repopulation of B cells following transplantation in vivo, were not transformable by Abelson murine leukemia virus, and had the physical properties of immature cells (Dorshkind and Phillips 1982, 1983).

The delayed kinetics of repopulation of B cells following bone marrow transplantation was demonstrated by comparing reconstitution of CFU-B in irradiated CBA/N mice grafted with cultured cells or fresh marrow. When grafted with the latter population, B-cell colonies are detected, usually by 1 week and consistently by 2 weeks, after reconstitution in the spleen. No CFU-B could be detected after 1 week in mice that received MBMC cells, and detection of B cells after 2 weeks was rare. The delayed kinetics was not due to the presence of fewer precursors in MBMC, since transplantation of even four times as many cultured cells as fresh marrow cells did not change the kinetics (Dorshkind and Phillips 1983). This lag in kinetics was confirmed by Nishikawa et al. (1985). These investigators further demonstrated that B-cell precursors in MBMC first lodge in the marrow after transplantation, and that their differentiated progeny subsequently migrate to the spleen. The inability of Abelson murine leukemia virus (A-MuLV) to transform MBMC cells also provided evidence that the B-cell precursors in the cultures were primitive cells. A-MuLV transforms large, immature bone marrow cells, and progenitors in the B-cell lineage have been shown to be particularly susceptible (Rosenberg and Baltimore 1980). The presence of A-MuLV target cells in BALB/c MBMC was examined at several times after recharge, and the findings demonstrated that transformable cells disappeared after 3 days. No target cells were observed in either nonadherent or adherent fractions. These findings appear contradictory to those of Teich et al. (1979), who reported that A-MuLV target cells were present in MBMC. However, it is important to note that the latter investigators infected cultures shortly after recharge with fresh bone marrow cells. It is not surprising that transformable cells were present at that time, since the cellular composition of recently established cultures resembles that of fresh bone marrow, a tissue with a high frequency of target cells. Finally, characterization of the MBMC B-cell precursor by velocity sedimentation demonstrated it to be a large cell with a rapid sedimentation rate, distinguishing it from more mature, small lymphocyte populations.

2.2 Lymphoid Bone Marrow Cultures

An alternative method for establishing long-term bone marrow cultures that resulted in the sustained production of B lymphocytes and their precursors was described by Whitlock and Witte (1982). These LBMCs are initiated by culturing a bone marrow cell suspension at 37° C in RPMI-1640 supplemented with 5% fetal calf serum and 2-mercaptoethanol. Steroids are not included in these cultures and are inhibitory. Under these conditions a supporting adherent layer establishes upon which long-term lymphopoiesis is dependent. The cultures are never recharged, and three characteristic phases of growth are observed. In the first the adherent layer establishes, and nonadherent cells, up to 20% of which express surface Ig, are present. This is followed by a second or crisis phase in which active hemopoiesis ceases. Following a variable period of one to several weeks, foci of adherent hemopoietic cells reappear and nonadherent lymphoid populations are present. The adherent foci are not as large as those in MBMC and generally contain fewer cells (Fig. 2). Nonadherent cells in LBMC have the morphologic appearance of lymphocytes (Fig. 2b).

All available evidence indicates that LBMCs are restricted to the growth of cells in the lymphoid differentiative pathway, since CFU-S and CFU-GM are not detected (PHILLIPS et al. 1984; KURLAND et al. 1984). It is possible that a pluripotent hemopoietic stem cell not detected by these assays (PHILLIPS 1985) is present and would produce myeloid progeny under appropriate conditions. However, this is unlikely, since lymphopoiesis ceases and myelopoiesis does not establish upon transfer of established LBMCs to MBMC conditions (K. DORSHKIND, unpublished observation).

However, B-cell precursors are present in the cultures. One indication of this is that cultured cell populations are susceptible to transformation by A-MuLV (WHITLOCK and WITTE 1982). A direct examination of the maturational state of the lymphoid cells in LBMC was made by the examination of Ig genes in nontransformed clonal lines that were developed by culturing limiting dilution numbers of nonadherent cells onto preestablished bone marrow feeder layers. Of the limited number of clones analyzed, two had both heavy- and light-chain gene rearrangements and three had only heavy-chain gene rearrangements. A sixth clone that appeared to represent a very early pre-B-cell stage was also isolated. This line grew too slowly to provide sufficient numbers of cells for direct analysis. However, examination of A-MuLV transformants of it indicated that 90% of these expressed the immature lymphoid marker terminal deoxynucleotidyl transferase (WHITLOCK et al. 1983).

Five of the above clones synthesized μ heavy-chain genes that comigrated on two-dimensional gels, suggesting that they represented the progeny of a single precursor cell that had rearranged its heavy-chain genes. That the clones were sibs of one another is not surprising, since they were isolated from cultures 12 weeks after initiation. Ig heterogeneity decreases as cultures reach this age, and the resultant Ig pauciclonality would suggest that the number of distinct stem/progenitor populations becomes limiting. However, analysis of Ig molecules synthesized by nonclonal populations of cells from LBMCs less than 3 months of age indicates a pattern of heterogeneity comparable to that of normal spleen cells.

Lymphocytes in LBMC often differentiate to the stage of expression of surface Ig. However, they remain more immature than those in fresh bone marrow, since no response to the B-cell mitogens dextran sulfate and LPS can be obtained in vitro (NAGASAWA et al. 1985). Nevertheless, the cultured cells do have the potential for normal lymphocyte differentiation after transplantation in vivo. KURLAND et al. (1984) demonstrated that cells from BALB/c cultures could reconstitute CFU-B in the spleen of F_1 progeny of BALB/c × CBA/N mice crosses, and could restore the ability to respond to the T-independent antigen TNP-Ficoll. Similarly, NAGASAWA et al. (1985) recovered B cells expressing surface Ig that could respond to LPS from the spleen of LBMC-reconstituted

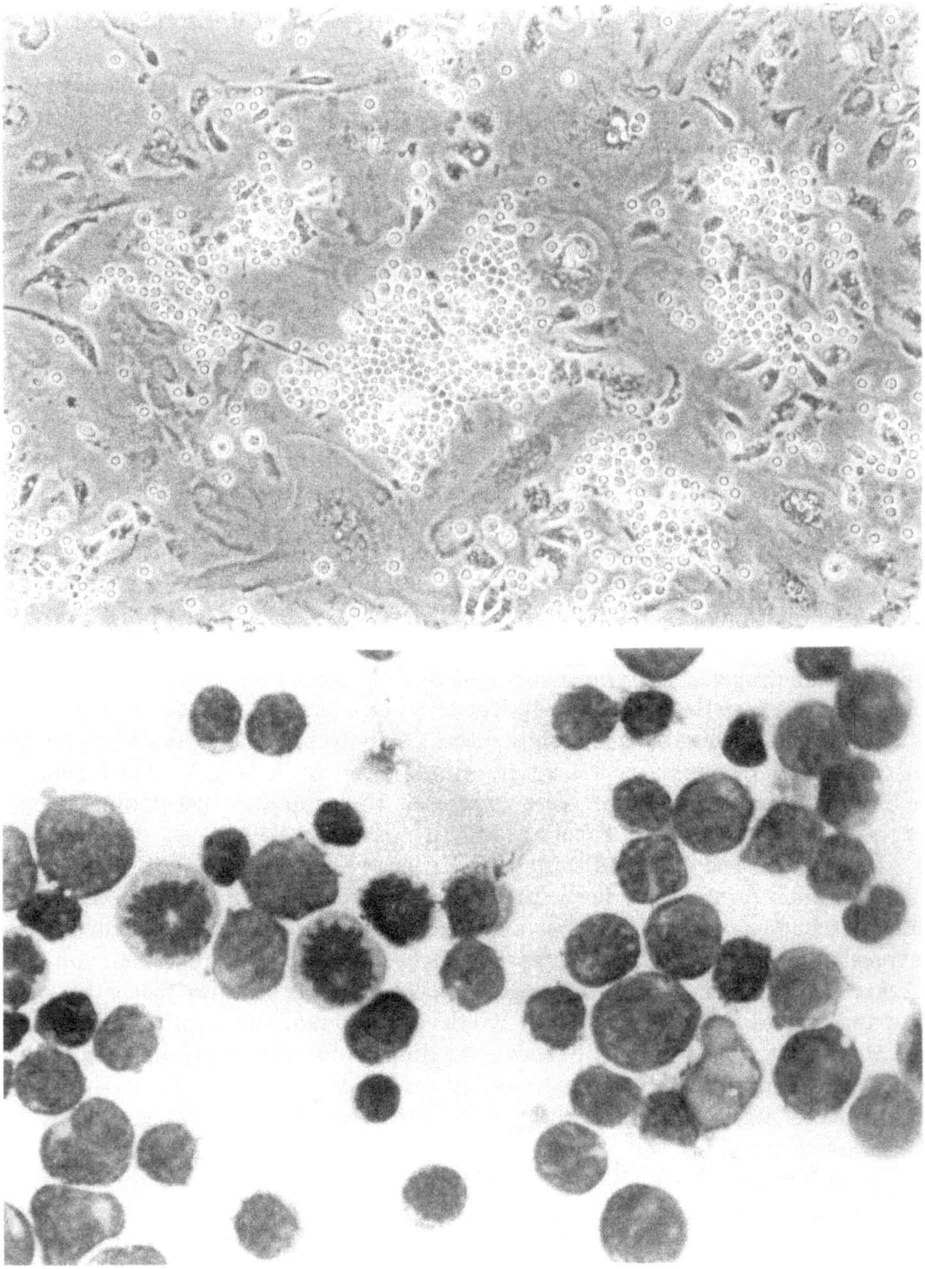

a

b

Fig. 2. **a** Phase-contrast micrograph of lymphoid bone marrow culture. Several foci of loosely arranged cells are present. **b** Nonadherent cells from lymphoid bone marrow culture. Cells are small to medium-sized lymphocytes. (From DORSHKIND et al. 1985)

mice. These findings suggest that a regulatory element necessary for the terminal stages of B-cell differentiation to proceed is absent from the cultures.

The use of normal or partially immunodeficient mice in the above studies made it difficult to distinguish the contribution of donor cells from that of radioresistant host elements, and to investigate the total lymphoid differentiative potential of the cultured cells. A collaborative effort between our two laboratories has circumvented these limitations by using a newly discovered severe combined immunodeficient (SCID) mouse as the focal point of a combined in vitro–in vivo system for the analysis of B-cell development. SCID mice are severely deficient in T and B cells and do not express serum Ig, but provide a normal microenvironment for lymphoid differentiation (Bosma et al. 1983; Dorshkind et al. 1984).

Preliminary experiments have shown that cells expressing surface Ig and CFU-B can be reconstituted in the spleen of SCID mice that receive a graft of LBMC cells, and that these B cells can secrete heterogeneous populations of IgM and IgG molecules. A particularly interesting series of results indicated that LBMC cells also reconstituted T-cell function in SCID mice. The ability of B cells to secrete multiple Ig isotypes, particularly IgG_{2a} and IgG_{2b}, is preferentially enhanced by the presence of T cells (Mongini et al. 1983), and both molecules were present in the serum of reconstituted mice. This was consistent with the fact that splenocytes from reconstituted mice could occasionally generate a cytotoxic response to alloantigens, and LBMC-engrafted mice always responded to immunization with the T-dependent antigen TNP-BSA. Thus, these data indicate that LBMC may provide an in vitro system for the analysis of T-lineage cells as well as those in the B-cell lineage.

Additional studies that define the maturational state of the lymphoid precursors in LBMC need to be conducted. For example, it is unclear whether the T-cell activity is due to mature T cells that contaminated the marrow inoculum used to establish the culture, or to a culture-derived prothymocyte. It is also unknown whether B-cell production is dependent on putative lymphoid stem cells or on B-cell-restricted precursors. In spite of these questions, the cultures are already proving to be a valuable tool in the analysis of molecular events that occur at various stages of B-cell development. For example, Whitlock et al. (1983) demonstrated that the various B and pre-B clonal sublines described above were all transformable by A-MuLV, indicating that the B-lymphoid target cells for that retrovirus are not restricted to a single developmental stage. Further analysis revealed that the progeny of a single transformed cell could undergo light-chain rearrangements and express those proteins. In a subsequent, related study Ziegler et al. (1984) analyzed κ-chain diversity among the progeny of an A-MuLV-transformed line obtained from the cultures, and showed that a single pre-B-cell clone could select a κ variable region from more than one family.

2.3 Fetal Liver Cultures

Murine fetal liver is a site of transient B lymphopoiesis, and studies of cells from the tissue have contributed to the understanding of B-cell development

(Melchers 1977; Owen et al. 1974; Paige et al. 1984). An in vitro fetal liver culture (FLC) system developed by Denis et al. (1984) allows the lymphoid elements in that tissue to grow and develop while maintaining traits characteristic of cells at that stage of development. These cultures are initiated by plating 14–17-day fetal liver onto preestablished bone marrow adherent layers. At the end of 1 week, cell growth initiates and nonadherent cells are transferred to fresh feeder layers. From this point on, cultures are maintained according to the protocols for LBMC (Whitlock et al. 1984). No distinct crisis phase is observed, although marked increases in cell production do not occur until 3–5 weeks after transfer to the secondary feeder layer. Cultures have continued to produce cells for at least 4 months. The appearance of the cultures is identical to that of LBMC, and small foci of loosely arranged cells are a characteristic feature. Nonadherent cells have the morphology of lymphocytes.

A detailed investigation of the cell lineages present in FLC has not been made, although cells from the cultures do not generate CFU-S or express the Mac-1 surface antigen. However, phenotypically identifiable pre-B cells are present, as up to 17% of the cells in 11-week-old cultures express cytoplasmic Ig. Only 1% of the cells in the original fetal liver inoculum bore cytoplasmic Ig, indicating expansion and/or selective survival of these immature cells in culture. Less than 1% of cells in FLC expressed surface Ig, in contrast to those from LBMC in which 30%–70% of the cells with cytoplasmic Ig expressed membrane Ig.

Southern-blot analysis of DNA from the cultured fetal liver cells was performed in order to examine Ig genes, and the findings further confirmed the presence of B-cell precursors in the cultures. Some nonadherent cells from FLC retained Ig genes in the germ-line configuration, and at least one chromosome remained in the unrearranged state for 3 months or longer at the heavy-chain locus in a subpopulation of cells in vitro. These cell populations could represent an early precursor population. Also present were other cells that had rearrangements of the heavy-chain genes. The number of such rearrangements decreased as the cultures aged, paralleling the shift to pauciclonality noted for LBMC. No dominant κ light-chain gene rearrangement was detected in any of the cells from the mass cultures.

The phenotypic and molecular characteristics of distinct cell populations in FLC were assessed by analyzing clonal lines isolated by A-MuLV transformation. Eight of 42 such transformants synthesized μ heavy chains. None of the transformants synthesized κ light chains. A-MuLV transformants from LBMC had three times the number of μ-producing cells. This phenotypic profile is characteristic of A-MuLV transformants from fresh fetal liver, indicating that FLC cells retain the characteristics of that tissue. All of the transformed cell lines analyzed had rearranged μ heavy-chain genes on both chromosomes. In six clones κ light-chain genes were examined and five of these had one rearranged and one germ-line gene. One line had both κ-chain genes in the germ-line configuration.

Preliminary evidence indicates that FLC cells can reconstitute B cells in SCID mice (K.A. Denis et al., unpublished observation). In contrast to reconstitution with LBMC cells, FLC cells can not confer the ability to respond to

T-dependent antigens, suggesting that the cultures maintain precursors with a more restricted developmental potential. Whether the B-cell precursors present in these cultures are the same as the early progenitor detected in the agar assay described by PAIGE (1983; PAIGE et al. 1984) remains to be determined.

3 Identification of Stromal Cells in Long-Term Cultures

Exogenous stimulatory elements potentially play a role in primary B-cell production (FULOP and OSMOND 1983), and the cells that form the adherent layer in vitro and provide a supporting framework for hemopoiesis in the marrow intersinusoidal space are a prime candidate population for the source of such external stimuli. These cells are referred to as stromal cells (LICHTMAN 1981), and they have been implicated in forming a hemopoietic microenvironment that regulates blood-cell development (WOLF and TRENTIN 1968). Stromal cells could mediate their regulatory effects by direct cell contact or by the production of soluble growth factors, and the data presented below suggest that both mechanisms may be operative. Since hemopoiesis in long-term cultures is dependent upon these populations, the various culture systems present the opportunity to analyze potential stromal cell–hemopoietic cell interactions. This section will review current knowledge regarding B-cell microenvironments in the cultures.

3.1 Myeloid Bone Marrow Cultures

A number of cell types, including fibroblasts, epithelioid cells, endothelial cells, and fat cells, have been described as components of the multilayered adherent layer in MBMC (Fig. 3; ALLEN and DEXTER 1976; TAVASSOLI 1982; DORSHKIND et al. 1985), and cellular interactions between these and developing blood cells have been observed. However, these morphologic analyses have limitations, and there is controversy about the exact composition of the adherent layer.

More detailed characterization of cells from the adherent layer has resulted from the development of stromal cell lines from cultures similar to those produced by Dexter (LANOTTE et al. 1982; SONG and QUESENBERRY 1984; ZIPORI et al. 1985; SONG et al. 1985). A complete discussion of the properties of all of these lines is beyond the scope of the present review, but one of them, the TC-1 line (SONG et al. 1985), is of particular relevance.

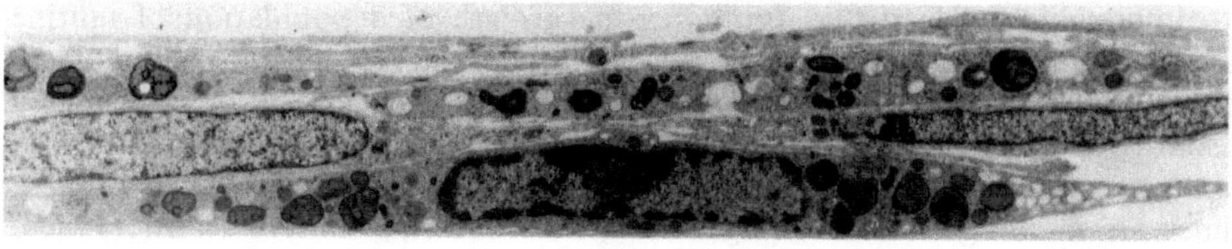

Fig. 3. Transmission electron micrograph of adherent layer of myeloid bone marrow culture. The multilayered arrangement of the stromal cells is apparent. ×12000. (From DORSHKIND et al. 1985)

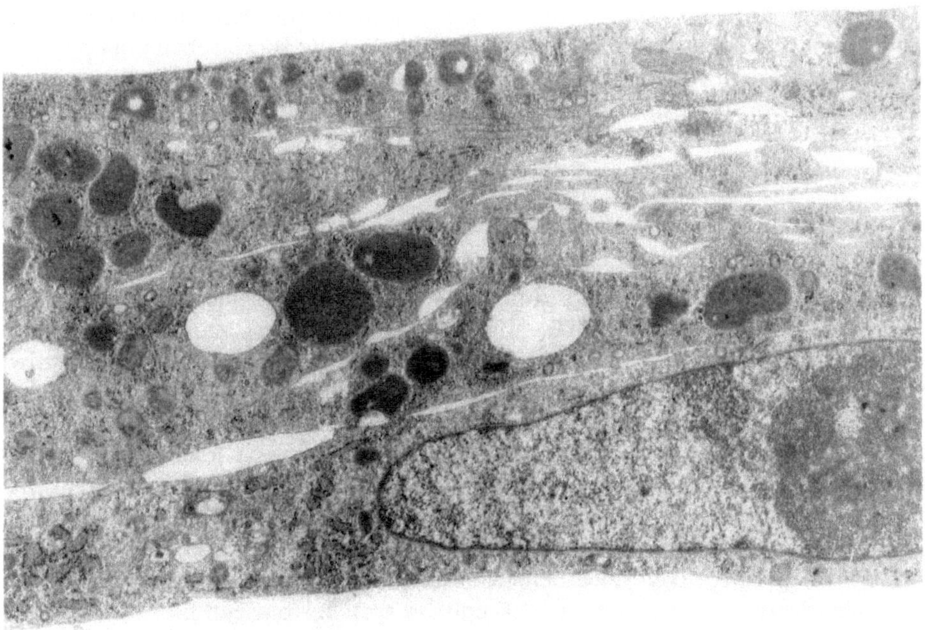

Fig. 4. Transmission electron micrograph of adherent layer of lymphoid bone marrow culture. The multilayered arrangement of the stromal cells is shown. ×12000. (From DORSHKIND et al. 1985)

TC-1 is a cell line expressing alkaline phosphatase that grows as an epithelioid monolayer and is capable of providing anchorage support for hemopoietic cells. A recent observation is that the cells secrete a factor(s) that stimulates marrow B-cell precursors isolated from fresh bone marrow to differentiate in vitro into cells that express cytoplasmic Ig (LANDRETH and QUESENBERRY, personal communication). The presence of a MBMC adherent-cell population with B-cell maturation activity suggests that the absence of B-cell microenvironments is not a reason for the failure of B lymphopoiesis in the cultures (see Sect. 3.4).

Whether the factor(s) secreted by TC-1 is the same as the B-cell differentiation factor present in the urine of patients with cyclic neutropenia (LANDRETH et al. 1985), or in the serum of New Zealand Black (NZB) mice (JYONOUCHI and KINCADE 1984), is unknown. The cyclic neutropenia factor is distinct from growth factors such as interleukins-1 and 2 (IL-1, IL-2), and B-cell growth factor (BCGF) that act on mature B cells, and from the activity attributed to interleukin-3 (IL-3) that promotes the growth of a population of murine B-cell precursors (PALACIOS et al. 1984; K.S. LANDRETH, personal communication). PALACIOS et al. (1984) have reported that IL-3-dependent clones isolated from fresh bone marrow of both normal and autoimmune mice display B-cell antigens such as the B-220 marker. Both heavy- and light-chain genes in these lines are in the germ-line configuration, but following in vivo transfer, these cells could be induced to differentiate and secrete antibody (PALACIOS and STEINMETZ 1985). This IL-3 activity must be distinguished from potential regulatory

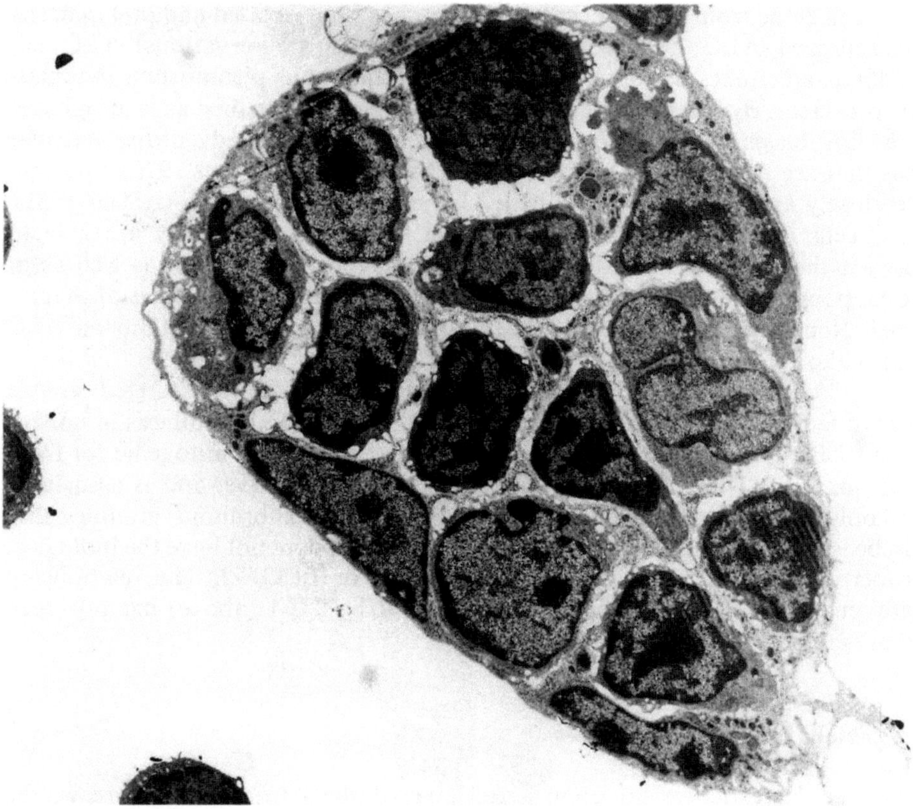

Fig. 5. Large stromal cell from lymphoid bone marrow culture. Numerous lymphocytes are shown associated with the cell. ×2500. (From DORSHKIND et al. 1985)

factors present in LBMC, as there is no evidence for that factor being a critical parameter in the cultures (O. WITTE, unpublished observation).

3.2 Lymphoid Bone Marrow Cultures

Since B-cell production occurs in LBMCs, analysis of the adherent layer would be expected to provide information about interactions between the stromal and developing B cells. A morphologic study demonstrated that the stromal cells in LBMC also grow in a multilayered arrangement (Fig. 4). Direct contact of developing B cells with stromal cells is evidenced by the observation that membrane infoldings of large stromal cells that measure up to 40 μm in diameter, surround developing lymphocytes (Fig. 5; DORSHKIND et al. 1985). Assuming that these cells have an in vivo counterpart, their characterization would provide important information regarding cells that form the B-cell microenvironment.

A cell clone from a LBMC adherent layer has been isolated and preliminarily characterized (ALC; D. ROBERTSON et al., unpublished observations). ALC cells are large, adherent, mononuclear, and exhibit numerous pleomorphic cytoplasmic processes. By several criteria, these cells can not be classified as macrophages or as fibroblasts. However, ALC cells do strongly resemble adventitial–reticular cells that are found within normal rodent bone marrow (WEISS 1970), and that are closely associated in situ with developing hemopoietic cells (ALLEN 1981). ALC cells apparently retain this function in culture, since they are able to support the continued proliferation of nonadherent LBMC cells as well as of clonal populations of pre-B or B cells, but not of splenic lymphocytes of thymocytes. Nonadherent cells from MBMC are also supported long term on ALC cells without loss of CFU-S activity.

ALC cells mediate this growth effect, at least in part, by secreted soluble growth factors. ALC culture supernatants will induce DNA synthesis in nonadherent LBMC and MBMC target cells, and are also highly mitogenic for fresh bone marrow. The ALC growth factor is protease sensitive, and is associated not only with culture supernatants but also with membranous granules that can be isolated from ALC cells. This soluble product does not have the biological properties of IL-1, IL-2, IL-3, or of BCGF-1 or BCGF-2. The relationship between ALC growth factor and that secreted by TC-1 (above) has not been investigated.

3.3 Fetal Liver Cultures

Since FLC are dependent upon a cell layer derived from bone marrow, the stromal cell–lymphoid cell interactions would presumably be the same as in LBMC. It is important to note, however, that adherent regulatory cells in fetal liver are capable of supporting B-cell differentiation (PAIGE 1983; PAIGE et al. 1984) from primitive fetal liver cell precursors.

3.4 Comparative Analysis of Stromal Cells in Different Long-Term Cultures

The presence of a MBMC stromal cell that can stimulate B-cell precursors to synthesize Ig suggests that the culture conditions are a key obstacle preventing the initiation of B lymphopoiesis. This is consistent with the observation that steroids are toxic to mature lymphocytes (PHILLIPS 1980) and that their addition to LBMC results in a cessation of cell production (WHITLOCK and WITTE 1982). However, the possibility that the conditions of MBMC prevent the establishment of all stromal cell populations necessary for the support of B-cell differentiation cannot be excluded. Thus, the stroma in MBMC may constitute a microenvironment that can only support myelopoiesis. A related question is whether the stroma in LBMC represents a lymphoid-restricted environment. This has been tested by generating purified populations of MBMC of LBMC stromal cells and measuring their hemopoietic support capacity.

Purified populations of stromal cells were prepared by treating established cultures with the antibiotic mycophenolic acid (MPA). This blocks inosine monophosphate dehydrogenase, an enzyme in the guanosine monophosphate synthesis pathway (Cass et al. 1977). Dividing hemopoietic cells are most susceptible to its action, while the adherent stroma, nondividing due to contact inhibitions, is not (Dorshkind et al. 1986). MPA-treated MBMCs or LBMCs were reseeded with bone marrow cells that had been passed through nylon wool, a procedure that yields a relatively purified population of hemopoietic cells uncontaminated by stroma; they were then shifted to LBMC or MBMC conditions respectively. Stromal cells established under MBMC conditions could support long-term B lymphopoiesis under LBMC conditions and LBMC stroma could support myelopoiesis under MBMC conditions. Thus, the different culture conditions do not result in the establishment of lymphoid- or myeloid-restricted microenvironments (Dorshkind 1986).

4 Induction of B Lymphopoiesis from Stem Cells in Vitro

The ability to induce the early B-cell precursor in MBMCs to differentiate and produce B cells would permit events that regulate B-cell production to be studied from its earliest stages. The fact that MBMC stroma can support B lymphopoiesis indicates that the major obstacle in attaining this goal is in fact the culture conditions. Our laboratories have independently investigated this and have shown that alteration of MBMC conditions results in B-cell production (Denis and Witte 1986; Dorshkind 1986).

Denis and Witte (1986) demonstrated that the transfer of nonadherent and loosely adherent cells from MBMC onto preestablished LBMC stroma resulted in the production of morphologically identifiable lymphoid cells within 7 days. The number of cells with myeloid morphology and surface markers decreased greatly during this time, correlating with a 90% death of nonadherent cells. Cells expressing cytoplasmic μ-chains were first observed 9–10 days after culture shift. Four to five weeks after culture shift more than 95% of the nonadherent cells present were lymphoid in morphology and 20%–40% expressed cytoplasmic μ-chains. The Ig heavy and light chains synthesized by these B-lymphocytes showed heterogeneity comparable to that of chains synthesized by the B-lymphocytes of conventional LBMC, when analyzed by two-dimensional gel electrophoresis.

The study by Dorshkind (1986) demonstrated that B lymphopoiesis could be induced by transferring a recharged MBMC to LBMC conditions, further confirming that the stroma in MBMC can support B-lymphopoiesis. Three weeks after this transfer, granulocyte–macrophage progenitors declined to undetectable levels. Cells that expressed the 14.8 surface antigen were present 1 week after, and cells expressing surface Ig 3 weeks after, the switch in culture conditions. This correlated with the appearance of CFU-B. The shift to lymphopoiesis could be observed morphologically; granulocytes and macrophages disappeared from the cultures by week 4, at which time almost all the cells had a characteristic lymphoid morphology. The transfer to LBMC conditions was not conductive to the survival of myeloid precursors, since switching the cultures back to MBMC conditions resulted in only low levels of transient myelopoiesis. Studies using these "switch" cultures will be important for the study of primary B-cell differentiation, since they are the first studies that demonstrate that early differentiation of B-lineage cells from stem-like cells in the bone marrow can occur in vitro.

5 Future Studies

Characterization of the maturational state of B-cell precursors in the various culture systems and identification of key elements of the hemopoietic microenvironment are important goals of future research using long-term culture systems. The cultures are also a useful physiologic system for investigating the expression and function of foreign genes transferred into hemopoietic populations by retroviral vectors (JOYNER et al. 1983; WILLIAMS et al. 1984; BOETTIGER et al. 1985). This technology also offers a means of identifying the progeny of selected hemopoietic precursors using the proviral integration site as a clonal marker (DICK et al. 1985; Van der PUTTEN 1985; KELLER et al. 1985).

Acknowledgements. The original work cited in this review was supported by grants to Kenneth Dorshkind from the National Institutes of Health (AI21256) and the Cancer Research Coordinating Committee and Academic Senate of the University of California, and to Owen N. Witte from the National Cancer Institute and the March of Dimes. Owen N. Witte is a faculty scholar of the American Cancer Society.

References

Abney ER, Cooper MD, Kearney JF, Lawton AR, Parkhouse RME (1978) Sequential expression of immunoglobulin in developing mouse B lymphocytes: a systematic survey that suggests a model for the generation of immunoglobulin isotype diversity. J Immunol 120:2041–2049

Allen TD (1981) Hemopoietic microenvironments in vitro: ultrastructural aspects in microenvironments in haemopoietic and lymphoid differentiation. CIBA Foundation Vol. 84. Pitman Medical London, pp 38–67

Allen TD, Dexter TM (1976) Cellular interrelationships during in vitro granulopoiesis. Differentiation 6:191–194

Aspinall R, Owen JJT (1983) An investigation into the B lymphopoietic capacity of long-term bone marrow cultures. Immunology 48:9–15

Boettiger D, Anderson S, Dexter TM (1985) Effect of SRC on hemopoietic progenitor cells: increased self-renewal without leukemia. Cell 36:763–773

Bosma GC, Custer RP, Bosma MJ (1983) A severe combined immunodeficiency mutation in the mouse. Nature 301:527–530

Cass CE, Lowe JK, Machak JM, Henderson JF (1977) Biological effects of inhibition of guanine nucleotide synthesis by mycophenolic acid in cultured neurblastoma cells. Cancer Res 37:3314–3320

Denis KA, Witte ON (1986) In vitro development of B lymphocyte from long-term cultured precursor cells. Proc Natl Acad Sci USA (to be published)

Denis KA, Treiman LJ, St Claire JI, Witte ON (1984) Long-term cultures of murine fetal liver retain very early B lymphoid phenotype. J Exp Med 160:1087–1101
Dexter TM, Spooncer E (1978) Loss of immunoreactivity in long-term bone marrow culture. Nature 275:135–136
Dexter TM, Allen TD, Lajtha LG (1977a) Conditions controlling the proliferation of haemopoietic stem cells in vitro. J Cell Physiol 91:335–344
Dexter TM, Moore MAS, Sheridan APC (1977b) Maintenance of hemopoietic stem cells and production of differentiated progeny in allogeneic and semiallogeneic bone marrow chimeras in vitro. J Exp Med 145:1612–1616
Dexter TM, Wright EG, Krizsa F, Lajtha LG (1977c) Regulation of haemopoetic stem cell proliferation in long-term bone marrow cultures. Biomedicine 27:344–349
Dick JE, Magli MC, Huszar D, Phillips RA, Bernstein A (1985) Introduction of a selectable gene into primitive stem cells capable of long-term reconstitution of the hemopoietic system of W/W mice. Cell 42:71–79
Dorshkind K (1986) In vitro differentiation of B lymphocytes from primitive hemopoietic precursors present in long-term bone marrow cultures. J Immunol 136:422–429
Dorshkind K, Phillips RA (1982) Maturational state of lymphoid cells in long-term bone marrow cultures. J Immunol 129:2444–2450
Dorshkind K, Phillips RA (1983) Characterization of early B lymphocyte precursors present in long-term bone marrow cultures. J Immunol 131:2240–2245
Dorshkind K, Keller GM, Phillips RA, Miller RG, Bosma GC, O'Toole M, Bosma MJ (1984) Functional status of cells from lymphomyeloid tissues in mice with severe combined immunodeficiency disease. J Immunol 132:1804–1808
Dorshkind K, Schouest L, Fletcher WH (1985) Morphologic analysis of long-term bone marrow cultures that support B lymphopoiesis or myelopoiesis. Cell Tissue Res 239:375–382
Dorshkind K, Johnson A, Collins L, Keller GM, Philips RA (1986) Generation of purified stromal cell cultures that support lymphoid and myeloid precursors. J Immunol Methods 89:37–47
Fulop GM, Osmond DG (1983) Regulation of bone marrow lymphocyte production. III. Increased production of B and non-B lymphocytes after administering systemic antigens. Cell Immunol 75:80–90
Johnson A, Dorshkind K (1986) Stromal cells in myeloid and lymphoid long-term bone marrow cultures can support multiple hemopoietic lineages and modulate their production of hemopoietic growth factors. Blood 68:1348–1354
Jones-Villeneuve E, Phillips RA (1980) Potentials for lymphoid differentiation by cells from long-term cultures of bone marrow. Exp Hematol 8:65–76
Joyner A, Keller G, Phillips RA, Bernstein A (1983) Retrovirus mediated transfer of a bacterial gene into mouse hematopoietic progenitor cells. Nature 305:206–208
Jyonouchi H, Kincade PW (1984) Precocious and enhanced functional maturation of B lineage cells in New Zealand black mice during embryonic development. J Exp Med 159:1277–1282
Keller G, Paige C, Gilboa E, Wagner EF (1985) Expression of a foreign gene in myeloid and lymphoid cells derived from multipotent haematopoietic precursors. Nature 318:149–154
Keller GM, Johnson GR, Phillips RA (1983) Hemopoiesis in spleen and bone marrow cultures. J Cell Physiol 116:7–15
Kincade PW (1977) Defective colony formation by B lymphocytes from CBA/N and C3H/HeJ mice. J Exp Med 145:249–263
Kincade PW, Lee G, Paige CJ, Scheid MP (1981) Cellular interactions affecting the maturation of murine B lymphocyte precursors in vitro. J Immunol 127:255–260
Kurland JI, Ziegler SF, Witte ON (1984) Long-term cultured B lymphocytes and their precursors reconstitute the B-lymphocyte lineage in vivo. Proc Natl Acad Sci USA 81:7754–7558
Landreth KS, Rosse C, Clagett J (1981) Myelogenous production and maturation of B lymphocytes in the mouse. J Immunol 127:2027–2034
Landreth KS, Kincade PW, Lee G, Medlock ES (1983) Phenotypic and functional characterization of murine B lymphocyte precursors isolated from fetal and adult tissues. J Immunol 131:572–580
Landreth KS, Engelhard D, Bear MH, Kincade PW, Kapoor W, Good RA (1985) Regulation of human B lymphopoiesis: Effect of urinary activity associated with cyclic neutropenia. J Immunol 134:2305–2309
Lanotte M, Scott D, Dexter TM, Allen TD (1982) Clonal preadipocyte cell lines with different

phenotypes derived from murine marrow stroma. Factors influencing growth and adipogenesis in vitro. J Cell Physiol 111:177–186
Lau CY, Melchers F, Miller RG, Phillips RA (1979) In vitro differentiation of B lymphocytes from pre-B cells. J Immunol 122:1273–1277
Lichtman MA (1981) The ultrastructure of the hemopoietic environment of the marrow: a review. Exp Hematol 9:391–420
Mauch P, Greenberger JS, Botnick LE, Hannon EC, Hellman S (1980) Evidence for structured variation in self-renewal capacity within long-term bone marrow cultures. Proc Natl Acad Sci USA 77:2927–2930
Melchers F (1977) B lymphocyte development in fetal liver. I. Development of reactivities to B cell mitogens in vivo and in vitro. Eur J Immunol 7:476–481
Mongini PKA, Paul WE, Metcalf ES (1983) IgG subclass, IgE, and IgA anti-trinitrophenyl antibody production within trinitrophenyl-Ficoll-responsive B cell clones. Evidence in support of three distinct switching pathways. J Exp Med 157:69–85
Nagasawa R, Kanagawa O, Tittle TV, Chiller JM (1985) In vivo maturation of in vitro derived pre-B cells. J Immunol 135:965–972
Nishikawa S, Kina T, Amagai T, Katsura Y (1985) Early B cell precursors in long-term bone marrow culture: selective development in the bone marrow of irradiated recipients. Eur J Immunol 15:696–701
Osmond DG, Nossal GJV (1974) Differentiation of lymphocytes in mouse bone marrow. II. Kinetics of maturation and renewal of antiglobulin-binding cells studied by double labelling. Cell Immunol 13:132–145
Owen JJT, Cooper MD, Raff MC (1974) In vitro generation of B lymphocytes in mouse fetal liver; a mammalian bursal equivalent. Nature 249:361–363
Paige CJ (1983) Surface immunoglobulin-negative B-cell precursors detected by formation of antibody-secreting colonies in agar. Nature 302:711–713
Paige CJ, Gisler RH, McKearn JP, Iscove NN (1984) Differentiation of murine B cell precursors in agar culture. Frequency, surface marker analysis and requirements for growth of clonable pre-B cells. Eur J Immunol 17:979–987
Palacios R, Steinmetz M (1985) IL-3 dependent mouse clones that express B-220 surface antigen, contain Ig genes in germ-line configuration, and generate B lymphocytes. Cell 41:727–734
Palacios R, Henson G, Steinmetz M, McKearn J (1984) Interleukin 3 supports growth of mouse pre-B cells clones in vitro. Nature 309:126–134
Phillips RA (1980) Enhanced lymphoid and decreased myeloid reconstituting ability of stem cells from long-term cultures of mouse bone marrow. J Supramol Struct 14:77–83
Phillips RA (1985) Comparison of different assays for multipotent haemopoietic stem cells. In: Ford RJ, Maizel AL (eds) Mediators in cell growth and differentiation. Raven, New York, p 135
Phillips RA, Bosma M, Dorshkind K (1984) Reconstitution of immune-deficient mice with cells from long-term bone marrow cultures. In: Wright DG, Greenberger JS (eds), Long-term bone marrow culture. Liss, New York, p 309
Rahal MD, Osmond DG (1984) Heterogeneity of bone marrow lymphocytes: Radioautographic detection of pre-B cells bearing cytoplasmic μ chains and of B and T lymphocytes and characterization of null lymphoid cells. Cell Immunol 87:379–388
Rosenberg N, Baltimore D (1980) Abelson virus. In: Klein G (ed) Viral oncology, Raven, New York, p 187
Rosse C (1976) Small lymphocyte and transitional cell populations of the bone marrow; their role in the mediation of immune and hemopoietic progenitor cell functions. Int Rev Cytol 45:155–290
Scheid MP, Landreth KS, Tung JS, Kincade PW (1982) Preferential but nonexclusive expression of macromolecular antigens on B-lineage cells. Immunol Rev 69:141–159
Schrader JW, Schrader S (1978) In vitro studies on lymphocyte differentiation. I. Long-term in vitro culture of cells giving rise to functional lymphocytes in irradiated mice. J Exp Med 148:823–828
Schrader JW, Goldschneider I, Bollum FJ, Schrader S (1979) In vitro studies in lymphocyte differentiation. II. Generation of terminal deoxynucleotidyl transferase-positive cells in long-term cultures of mouse bone marrow. J Immunol 122:2337–2339

Song ZX, Quesenberry PJ (1984) Radioresistant murine marrow stromal cells: a morphologic and functional characterization. Exp Hematol 12:523–533

Song ZX, Shadduck RK, Innes DJ, Waheed A, Quesenberry PJ (1985) Hematopoietic factor production by a cell line (TC-1) derived from adherent murine marrow cells. Blood 66:273–281

Tavassoli M (1982) Ultrastructural and radiobiological characterization of stromal cells in continuous, long-term marrow cell line. In: Johari O (ed) Scanning electron mcirosocpy. SEM, AMF O'Hare, Chicago, p 349

Teich N, Boss M, Dexter TM (1979) Infection of mouse bone marrow cells with Abelson murine leukemia virus and establishment of producer cells. In: Neth R, Gallo RG, Hofscheider PH, Mannweiler K (eds) Hematology and blood transfusion. Springer, Berlin Heidelberg New York, p 487 (Modern trends in human leukemia)

Van der Putten H, Botteri FM, Miller AD, Rosenfeld MG, Fan H, Evans RM, Verma IM (1985) Efficient insertion of genes into the mouse germ-line via retroviral vectors. Proc Natl Acad Sci USA 82:6148–6152

Weiss L (1970) Transmural cellular passage in vascular sinuses of rat bone marrow. Blood 36:189–208

Whitlock CA, Witte ON (1982) Long-term culture of B lymphocytes and their precursors from murine bone marrow. Proc Natl Acad Sci USA 79:3608–3612

Whitlock CA, Ziegler SF, Treiman LJ, Stafford JI, Witte ON (1983) Differentiation of cloned populations of immature B cells after transformation with Abelson murine leukemia virus. Cell 32:903–911

Whitlock CA, Robertson D, Witte ON (1984) Murine B cell lymphopoiesis in long-term culture. J Immunol Methods 16:353–369

Williams N, Jackson H, Sheridan APC, Murphy MJ, Eisle A, Moore MAS (1978) Regulation of megakaryopoiesis in long-term murine bone marrow cultures. Blood 51:245–255

Williams DA, Lemischka IR, Nathans DG, Mulligan RC (1984) Induction of new genetic material into pluripotent haemopoietic stem cells of the mouse. Nature 310:476–480

Wolf NS, Trentin JJ (1968) Hemopoietic colony studies. V. Effect of hemopoietic organ stroma on differentiation. J Exp Med 127:205–214

Ziegler SF, Treiman LJ, Witte ON (1984) K gene diversity among the clonal progeny of pre-B lymphocytes. Proc Natl Acad Sci USA 81:1529–1533

Zipori D, Duskin D, Tamir M, Argaman A, Toledo J, Malik Z (1985) Cultured mouse marrow stromal cell lines. II. Distinct subtypes differing in morphology, collagen types, myelopoietic factors, and leukemic cell growth modulating activities. J Cell Physiol 122:81–90

Peyer's Patches and the Early Development of B Lymphocytes

J.D. REYNOLDS

1 Introduction

A number of organs contribute to the development of the B-lymphocyte lineage during ontogeny. In the fetal mouse, one of the earliest phases of B-cell development occurs in the liver, then the focus shifts to the spleen and subsequently to the bone marrow (see review by MELCHERS 1979). The possibility that Peyer's patches (PPs) also contribute to early B-cell ontogeny has been the topic of a number of studies, particularly in the late 1960s when the role of the bursa of Fabricius in the chicken was becoming clearer. It was argued that mammalian PPs and the avian bursa might have similar functions (see reviews by COOPER and LAWTON 1973; REYNOLDS et al. 1981). However, subsequent studies seemed to rule out this possibility because tissues such as the fetal liver (OWEN et al. 1974) and bone marrow (OSMOND and NOSSAL 1974) were found to be important sites of early B-cell development in some mammals. Furthermore, evidence was obtained against the involvement of PPs during the period of B-cell expansion in the neonatal mouse (FRIEDBERG and WEISSMAN 1974). Other studies established that PPs are sites where immune responses are initiated to antigen from the intestinal lumen. The stimulated B cells in PPs undergo terminal differentiation along a pathway that contributes to mucosal tissues being populated with plasma cells (CRAIG and CEBRA 1971; see reviews by MAYRHOFER 1984; STROBER and JACOBS 1985).

It is now considered that birds and mammals have evolved different ways of generating their complement of B cells. In birds the microenvironment re-

Department of Medical Physiology, University of Calgary, 3330 Hospital Drive NW, Calgary, Alberta, Canada

Current Topics in Microbiology and Immunology, Vol. 135

quired for early B-cell development seems to exist in the lymphoepithelial tissues of the bursa, whereas in mammals it seems to be provided by the liver and bone marrow. However, this generalization may not be valid in view of some recent studies in sheep, in which a lymphoepithelial organ, the PP, is the major source of B lymphocytes during early postnatal life.

The role of PPs as secondary lymphoid tissues contributing to terminal B-cell differentiation has been extensively studied and reviewed (MAYRHOFER 1984; STROBER and JACOBS 1985), and will therefore not be covered here. Emphasis will be placed on examining the hypothesis that PPs are also primary lymphoid tissues that influence the early phase of B-cell ontogeny. Support for this proposal comes from evidence of many important similarities between PPs in sheep and the avian bursa. Peyer's patches are major sites of B-cell production for export to other tissues and antigen does not have to be present for lymphopoiesis to occur.

2 Peyer's Patches as a Site of B-Cell Production

One function of the lymphopoietic zones in the PPs (Fig. 1) of all species that have been examined is to produce B lymphocytes. As many as 95% of all lymphocytes in the lymphopoietic zone have surface immunoglobulin (FRIEDBERG and WEISSMAN 1974; BUTCHER et al. 1982; LARSEN and LANDSVERK 1985). The exact proportion has not been determined because most B cells in the lymphopoietic zone have a low density of surface immunoglobulin (ROSE and MALCHIODI 1981; BUTCHER et al. 1982; MIYASAKA et al. 1984), making it difficult to classify the cells in situ using immunohistochemical techniques. Analysis is further complicated by the high concentration of free immunoglobulin that surrounds the cells (GROSSI et al. 1977; OPSTELTEN et al. 1982). Nevertheless, even with this limitation, it is clear that cells in the lymphopoietic zone are predominantly B cells and that T cells represent only a few percent of the total (HOFFMANN-FEZER et al. 1976; BUTCHER et al. 1982; MAYRHOFER et al. 1983; MIYASAKA et al. 1984).

In addition to being B cells, most cells in the lymphopoietic zone are proliferating (Fig. 1; MEUWISSEN et al. 1969; GULLIANI et al. 1974; REYNOLDS and MORRIS 1983). The rate of proliferation of cells in the PPs of lambs has been estimated from the rate at which the cells were arrested in metaphase after the injection of vincristine (REYNOLDS 1986). Examination of the lymphopoietic zone in histological sections of PPs showed that during the first 3 h after the initiation of metaphase arrest about 18% of the lymphocytes had entered metaphase (Fig. 2). From the rate of increase in the frequency of metaphase cells it was calculated that 5%–6% of the lymphocytes, which would be mostly B cells, divide each hour. Thus, the number of B cells in the lymphopoietic zone can double in about 20 h.

There are few other sites in the lamb with such a high proportion of proliferating lymphocytes. For example, in the thymic cortex cells divide at one-tenth of the rate found in the lymphopoietic zone of PP (J.D. REYNOLDS, in prepara-

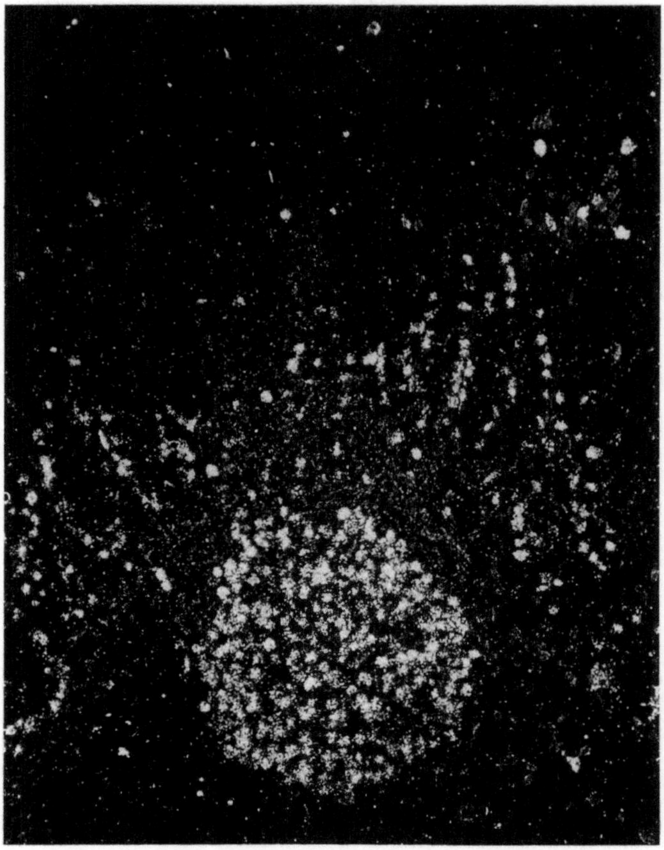

Fig. 1. An autoradiograph of a jejunal Peyer's patch from a fetal lamb after 140 days of gestation. Exposure to [^{3}H]-thymidine for 1 h in vivo is sufficient to label a high proportion of the cells in the follicle, confirming that it is a zone of intense lymphopoiesis. The dome region, which projects from the lymphopoietic zone towards the intestinal lumen, contains very few labeled cells. Also, lymphopoiesis is not a significant feature of the interfollicular region, which can be seen on either side of the follicle (× 200)

tion). Only the follicles in the chicken bursa, and germinal centers in the spleen and lymph nodes have a similar rate of lymphopoiesis (J.D. REYNOLDS, in preparation). Each is a site of B-cell lymphopoiesis, but there are reasons for believing that the lymphopoietic zone of the lamb PP is more like a bursal follicle than a germinal center (see Sect. 5).

3 Peyer's Patches as the Major Source of B Cells in Young Sheep

The lymphopoietic zone is clearly a site of rapid B-cell generation, but this does not justify the claim that PPs are major sites of B-cell production. FRIEDBERG and WEISSMAN (1974) examined neonatal mice for evidence that PPs might

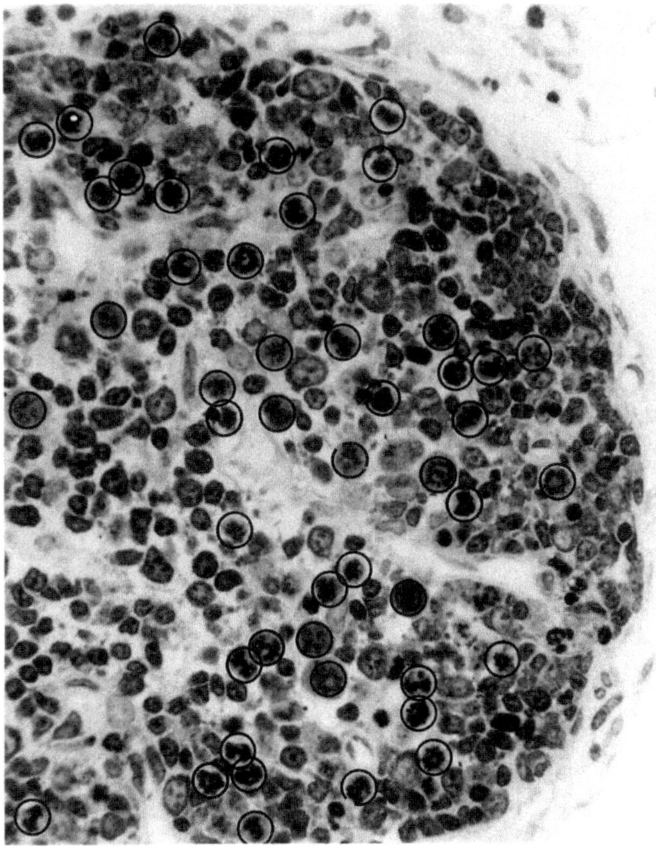

Fig. 2. The lymphopoietic zone from the ileocecal Peyer's patch of a 137-day-old fetal lamb, 3 h after the intravenous administration of vincristine. The *circles* show the location of the cells that had entered metaphase during this period. The high frequency of metaphase cells indicates the magnitude of lymphocyte production (×450)

make a major contribution to the postnatal expansion of the B-cell system. They showed that the number of B cells in the neonatal mouse increased logarithmically and that most of these cells were newly formed. Thus, the source of the B cells must be a site of rapid proliferation. Examination of PPs showed that fewer than 10% of the cells had surface immunoglobulin, and that only 35%–42% of these cells became labeled during a 1-h exposure in vivo to [^{3}H]-thymidine. It was clear that this population was far too small to account for the normal increase in the number of B cells in the neonatal mouse. However, this conclusion cannot be extended to all mammals. For example, in sheep, which is the only other species to be examined, there is clear evidence from several different approaches that PP lymphopoiesis can account for the postnatal expansion of B cells.

The PPs in a lamb contain about 100000 follicles, each 1–2 mm high (REYNOLDS and MORRIS 1983) and filled with dividing B cells (Fig. 1). The most spectacular feature of PPs in sheep, and certain other species, is the ileocecal PP that covers most of the circumference and length of the last 1–2 m of the small intestine. This region, together with the 30–40 other PPs, covers about 17% of the length of the small intestine and accounts for approximately 1% of the body weight of a 2-month-old lamb (REYNOLDS and MORRIS 1983).

The PPs in a 2-month-old lamb contain about 1.45×10^{11} lymphocytes, which is similar to the number of cells in the thymus (REYNOLDS 1986). No other organ in the lamb, including the bone marrow (M. MIYASAKA et al., unpublished observations) contains anywhere near this number of B cells (REYNOLDS 1986). The metaphase-arrest technique was used to determine that about 2.8% of the 1.45×10^{11} PP lymphocytes, or 4.1×10^9 new cells, are produced by cell division each hour. This is certainly an underestimate of the true value, because the preparation of cell suspensions during both the quantitation of PP cells and following intravenous vincristine administration would destroy many of the fragile larger cells, especially those arrested in metaphase (REYNOLDS 1986). Nevertheless, the method is valuable in that it defines the lower limit of lymphocyte production by all PPs in a lamb. Since most dividing cells are located in the follicles (Fig. 1), it is probable that the majority of the 3.6×10^9 cells produced in the PPs each hour are B cells.

How does the rate of cell production in the PPs of a lamb compare with the number of cells found in other organs? Most newly formed cells leaving the PPs travel via the lymphatics (REYNOLDS and PABST 1984). Therefore, about 4.1×10^9 cells, weighing approximately 3 g, should travel via the lymph from the intestine each hour and enter the mesenteric lymph nodes, which weigh approximately 30 g (REYNOLDS 1980). Therefore, every 10 h the mesenteric lymph nodes should receive their own weight in cells derived from lymphopoiesis in PP. The actual number of cells leaving the intestine in the lymph has been determined by cannulating the common intestinal lymphatic after excising the mesenteric lymph nodes (REYNOLDS 1986). The majority of these cells are long-lived recirculating cells and the number of newly formed cells coming from PPs was estimated to be about 1×10^8 to 2×10^8 per hour, or 2.5–5% of the number produced in the PPs per hour (REYNOLDS 1986). These and other investigations of production and emigration of newly formed cells from PPs (PABST and REYNOLDS 1986), question how a lamb could accommodate the large number of lymphocytes being produced. It was proposed that the vast majority of the B cells do not survive long enough to leave their site of formation in PPs. They are probably eliminated by the macrophages, found throughout the lymphopoietic zone, which contain the remnants of lymphocytes (REYNOLDS 1986). The production both of T cells in the thymus (SCOLLAY et al. 1980) and of B cells in PPs gives the impression that the mechanisms involved are highly wasteful. Perhaps by eliminating a large proportion of the newly formed cells, the PPs contribute in some crucial way to the screening and selection of B cells (see Sects. 6, 7).

4 The Influence of Peyer's Patches in Sheep Is Not Limited to the Gut

The current dogma is that PPs are sites where antigen enters from the gut lumen and initiates B-cell proliferation and differentiation. This involves blast transformation, emigration via the lymphatics, and localization in the gut lamina propria, followed by maturation into IgA-secreting plasma cells. The existence of this pathway is supported by a number of investigations (CRAIG and CEBRA 1971; HUSBAND and GOWANS 1978), including some in the sheep (PABST and REYNOLDS 1986). However, other studies in the sheep have shown that this is not the only fate of the cells produced in PPs.

Because about 90% of the PPs are concentrated in the last 1–2 m of the small intestine of a lamb (REYNOLDS and MORRIS 1983), it is possible to study the function of the gut-associated lymphoepithelial tissues in mammals by using surgical ablation techniques similar to those used in studying the function of the chicken bursa. The equivalent of bursectomy can be done in sheep by surgically excising the terminal 2 m of small intestine either before, or just after, birth when PPs are small. GERBER and colleagues (GERBER et al. 1986) were the first to show that ileectomy caused lambs to develop severe B-cell deficiencies. Subsequent studies showed that the deficiencies were not restricted to the gut-associated tissues but were found in all tissues examined, including lymph nodes remote from the gut, spleen, and blood (J.D. REYNOLDS, unpublished observations). It was also possible to abolish local B-cell lymphopoiesis by isolating the PPs from the intestinal tract (REYNOLDS and MORRIS 1984). If this was done near birth, before adult levels of B cells were established, the lambs developed widespread B-cell deficiencies (J.D. REYNOLDS, unpublished observations). The results observed after interfering with PP development are in accordance with the observation that PPs are sites of extravagant B-cell lymphopoiesis (see Sect. 3).

The nature and fate of cells leaving PPs in sheep has also been examined using extracorporeal perfusion to label selectively the ileocecal PP with fluorochromes or [^{3}H]-thymidine. This technique labeled most PPs in the lamb and established that most emigrant cells, including those produced locally, were small lymphocytes (REYNOLDS and PABST 1984; PABST and REYNOLDS 1986) with surface immunoglobulin (J.D. REYNOLDS and R. PABST, to be published). It was also shown that B cells from PPs migrated to all other lymphoid tissues (PABST and REYNOLDS 1986; J.D. REYNOLDS and R. PABST, to be published), as was predicted from the results of ileectomy experiments. For example, in the lymph nodes and spleen about 6% of the B cells came from PPs during a 24 h period (J.D. REYNOLDS and R. PABST, to be published). The newly produced cells that had migrated from PPs to the intestinal lamina propria included a high proportion of plasma cells, which contrasted with the non-gut tissues where PP-derived plasma cells were rare (PABST and REYNOLDS 1986). This result also supports the notion that antigen contributes to the events occurring in PPs after birth.

5 The Generation of B Cells in Peyer's Patches Is Independent of Antigen

When antigen enters PPs it apparently alters the fate of some of the resident B cells. It is also widely believed that this antigen stimulates B-cell lymphopoiesis. The evidence normally quoted to support this hypothesis is that PPs are poorly developed in mice maintained in an "antigen-free" environment after birth (POLLARD and SHARON 1970). There is also evidence from studies in sheep (REYNOLDS and MORRIS 1984), rabbit (PEREY and GOOD 1968), and mouse (FERGUSON and PARROTT 1972) that the lymphopoietic zone does not develop normally in PPs isolated from the intestinal tract. The requirement for contact with the intestinal tract is particularly evident during the postnatal period, when PPs grow rapidly. However, studies in the lamb showed that if PPs reached a mature size before being isolated from the intestinal tract, lymphopoiesis would continue at normal rates for at least 4 months (J.D. REYNOLDS et al., unpublished observations). There is also evidence that the same is true in the mouse (KAMINSKI 1976). Peyer's patches seem to be particularly sensitive to alterations in their environment while they are developing, but once developed they are less susceptible to change.

Even though PPs contain antigen-specific B cells that are proliferating, it cannot be assumed that the proliferation is caused by antigen entering from the gut lumen. The genesis of B cells in PPs is also particularly sensitive to other factors, including stress, the influence of the thymus, and probably also changes in blood flow. The PPs in both antigen-free animals and in isolated segments of gut develop under quite abnormal conditions in which several factors could contribute to the changes observed.

The evidence that B-cell genesis in PPs is independent of antigen comes from the examination of prenatal PP development in normal animals that had not been manipulated in any way. Peyer's patches develop before birth in man (CORNES 1965), cattle (CARLENS 1928; DOUGHRI et al. 1972), and sheep (CARLENS 1928; REYNOLDS and MORRIS 1983). In the fetal lamb it is also evident that PPs, like the thymus but unlike the spleen and lymph nodes, become histologically mature before birth. Each PP and each individual follicle in a PP is formed before birth, the process beginning about midway through gestation (REYNOLDS and MORRIS 1983). Proliferating cells can be seen in rudimentary PPs, and during the last month before birth the PPs are sites of active B-cell genesis (REYNOLDS 1981; REYNOLDS and MORRIS 1983). Furthermore, the metaphase-arrest technique has been used in sheep to show that cells in the lymphopoietic zone of PPs in the fetus enter mitosis at the same rate as cells in PPs after birth (REYNOLDS 1985).

The prenatal maturation of PPs in the lamb is significant because it occurs at a time when the animal is devoid of foreign antigenic material. The lamb becomes immunologically competent before birth (SILVERSTEIN et al. 1963; FAHEY and MORRIS 1978), but the placenta excludes exogenous antigen (BRAMBELL 1970) and ensures that the lymphoid apparatus, except for the thymus and the PP, appears dormant until after birth. It is notable in the fetal lamb that the mesenteric lymph nodes contain no germinal centers or other signs of reactivity to antigen. The lymph flowing to the mesenteric lymph nodes comes directly from the PPs, where B cells are dividing rapidly (REYNOLDS and MORRIS 1984); this seems to exclude the possibility that the gut contains some form of antigenic stimulus. In other experiments the injection of antigen into the fetal intestine was found to have no impact on either the histology of PPs or on the number of proliferating cells they contain (REYNOLDS and MORRIS 1984).

6 The Relationship Between the Development of Peyer's Patches and the B-Cell Lineage in Sheep

The observation that B-cell lymphopoiesis in the PPs of the developing sheep fetus is not related to antigen is the most compelling evidence in support of the contention that PPs contribute to early B-cell development. The establishment of B-cell lymphopoiesis in the chicken bursa also occurs without any contribution by conventional antigen (LYDYARD et al. 1976). One other important similarity between the avian bursa and the sheep ileocecal PP is that both involute (as does the thymus) at about the time the animal reaches sexual maturity (REYNOLDS and MORRIS 1983; REYNOLDS 1985). Both apparently contribute to the development of the immune system, and are then dispensed with once this role has been fulfilled. Thus, a closer examination of the relationship between PPs and early B-cell differentiation seems justified.

Regions in the jejunum that were presumed to be primordial PPs first appear on about the 68th day of the 150-day period of gestation. This time is marked by the appearance in the lamina propria of large basophilic cells (Fig. 3; REYNOLDS and MORRIS 1983) similar to those that first populate the rudiments of the thymus (MOORE and OWEN 1967) and bursa (ACKERMAN and KNOUFF 1959). Something must attract the basophilic cells into the intestinal lamina propria and keep them there. It is unlikely to be a random event since at least one PP (the ileocecal PP) always develops in the same location.

In the bursal anlage of the chick embryo, the lymphoid stem cells apparently mature within clusters of epithelial cells that form budlike projections from the mucosal epithelium (ACKERMAN and KNOUFF 1959). A similar structure does not develop in the sheep, although large basophilic cells are often associated with a dome-shaped bulge in the epithelium. Also in the lamina propria, adjacent to the basophilic cells, are reticular cells that form the framework of the developing PP (Fig. 3). These epithelial cells might have some inductive role, perhaps comparable to that of epithelial cells in the lymphoid–epithelial interaction in the thymus (see review by EWIJK 1984).

Based on what is known about the role of the liver in seeding the lymphoid organs in other species, it seems possible that the liver is the source of the lymphoid progenitors in sheep PP. The liver is a major site of hemopoiesis in the fetal sheep at a time in ontogeny when the thymus is the only lymphoid organ that has developed (AL SALAMI et al. 1985). The bone marrow develops at approximately the same time as the PPs and so it is less likely, though not excluded, as a source of PP-progenitors.

Nothing is known about the degree of differentiation of the large basophilic cells that first populate PP. They could be multipotential stem cells destined for the B-lymphocyte pathway under the influence of unknown stimuli generated within the microenvironment of the developing PP. Alternatively, the PP progenitor cells might be committed to the B-cell lineage, and perhaps have reached the status of a pre-B-lymphocyte, similar to those that have received so much attention in some other species. Pre-B-lymphocytes were not found after birth in either the PP-like tissues of the rabbit (MCELROY et al. 1981) or the lamb (J.D. REYNOLDS, unpublished observation); however, the possibility that the first cells to populate the PPs are pre-B cells or even B cells cannot be excluded.

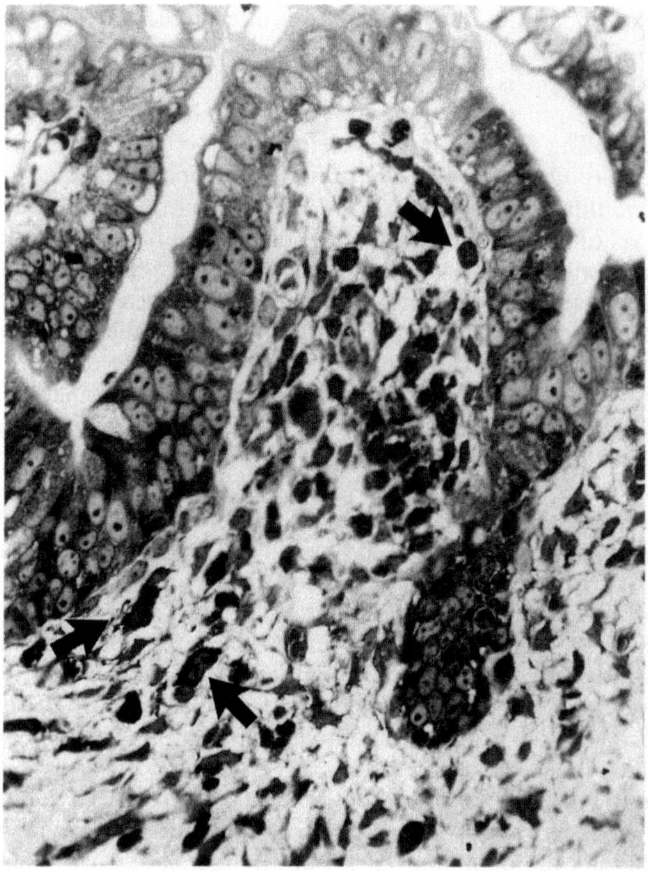

Fig. 3. A region presumed to be a primordial Peyer's patch in the jejunum of a 68-day-old fetal lamb. *Arrows* indicate some of the large, intensely staining cells that distinguish the region. Pale-staining reticular cells form the stroma of the tissue (× 500)

Analyses based on the techniques of immunohistochemistry or in situ hybridization should help resolve whether the immunoglobulin gene is being transcribed in the basophilic cells.

Peyer's patches develop in the fetal lamb at about the same time that B cells can first be detected in the blood and spleen. The next 10–20-day period in the growth of the fetus is a crucial time in the ontogeny of humoral immunity. During this period the lamb develops the capacity to mount a mature antibody response to a wide variety of antigens (Fahey and Morris 1978). Therefore, the antibody repertoire has undergone extensive diversification by this time. One important question still to be answered is whether the B cells in the 70- to 80-day-old fetus were nurtured in the PP.

The notion that PPs might be essential for early B-cell development was studied in the 1970s by Cooper and Dawes (cited by Cooper 1983). They excised the intestinal tract from fetal sheep after 55 days of gestation, then returned the fetus to the uterus. The fetuses contained no B cells at the time of surgery, but a few weeks later, in spite of the radical procedure, normal numbers of

B cells had developed. This was interpreted as evidence that PPs are not the sites of early B-cell development in the fetal lamb. However, an alternative explanation is that in the absence of PPs, other tissues take over as sites of early B-cell differentiation. A similar occurrence could also account for the surprising finding that bursectomy of the chicken at a very early stage of ontogeny (60 h in ovo) did not prevent the appearance of cells with surface immunoglobulin (JALKANEN et al. 1983). After hatching, the number of B cells and the level of immunoglobulin were either normal or only moderately reduced (GRANFORS et al. 1982; JALKANEN et al. 1983), yet these birds were only able to synthesize a severely restricted pattern of specific antibody following immunization (JALKANEN et al. 1984). The range of antibodies that a chicken could produce was also diminished following bursectomy at day 11 of incubation, but not if normal bursal function was allowed to continue until just before hatching (HUANG and DREYER 1978). These studies support the proposition that the bursa has an important impact on the antibody repertoire of the B cells.

Peyer's patches have a major role in the manufacture of an animal's complement of B cells, but do they influence the properties of these cells? No studies have been done which address this question, and so a number of possibilities exist. First, PPs might be sites of B-cell amplification. This view was originally proposed by Waksman and his colleagues after their studies of the rabbit appendix (reviewed by WAKSMAN and OZER 1976). They suggested that the amplification is antigen driven, although this is not the case in the midterm fetal lamb. Nevertheless, other aspects of the proposal might be true. For example, a cell that had undergone immunoglobulin gene rearrangement in an organ such as the liver might migrate to a PP where it would divide rapidly, giving rise to many identical progeny that could travel to other lymphoid tissues.

Secondly, this hypothesis could be modified to incorporate the observation that immunoglobulin genes, like all other genes, are susceptible to somatic mutation. In fact, regions of the variable gene mutate more frequently than most other genes (WABL et al. 1985). Consequently, the antigenic specificity of the B cells leaving PPs would probably differ from that of the progenitor cells that entered PP. The rate of cell division in PPs after 70–80 days of gestation, when the animal is immunocompetent, has not been determined, but by the last month of gestation all the B cells in the lymphopoietic zone divide, on average, once every 20 h (REYNOLDS 1985). Each time a B cell divides, and about 4.1×10^9 PP cells divide each hour, there is a high probability that a mutation will occur and that a functionally different B cell will be created. At a rate of about 10^{-5} mutations per nucleotide per cell division (WABL et al. 1985), a mutation in one of the 320 nucleotides of a variable gene could occur as frequently as once in every 300 cell divisions. Thus, the PPs might generate as many as 10^7 mutant B cells every hour.

7 Concluding Comments

A number of reasons for proposing that PPs are primary lymphoid organs in sheep have been outlined in this article. They are sites of extensive, antigen-independent, B-cell proliferation at a time in ontogeny when the humoral immune system is developing. Also, most PPs in sheep have stopped producing B cells by the time the animal is fully grown.

One important gap in our knowledge is the nature of the precursor cell that populates the primordial PP. However, even if the cell has already been committed to the B-cell lineage it is likely that PPs have other ways of influencing immunoglobulin diversity. JERNE'S hypothesis (1971), that primary lymphoid organs like the thymus and bursa are sites that foster mutations, and hence diversification of the immunological repertoire, could apply to the PPs of sheep. Any theory of how PPs might influence the B-cell repertoire must also take into account the high rate of cell death associated with PP lymphopoiesis (see Sect. 3). This is a fruitful topic for speculation but, as with the thymus, no satisfactory explanation exists. Lymphocyte death in PPs might be related to the elimination of self reactive cells, of non-functional mutations, or of cells that have not differentiated in the direction of IgA production, or even, as suggested by JERNE (1971) for the thymus, to the elimination of cells that have not mutated, thus fostering the selection and survival of mutant cells.

The hemopoietic function of the liver, spleen, and bone marrow is not restricted to the lymphoid lineage, as these organs also have a major role in the genesis of erythrocytes, granulocytes, monocytes, and platelets. The situation is clearly different in both sheep PPs and the avian bursa since the microenvironment of these tissues only fosters B-cell lymphopoiesis. These two lymphoepithelial tissues share so many important features that it seems reasonable to consider the possibility that they are functionally equivalent. Although the status of the bone marrow in the B-cell system of the sheep remains uncertain, enough is now known about PPs to indicate that these tissues, not the bone marrow, have the greatest impact on the number of B cells in sheep and, possibly, on the quality of the B cells. Therefore, the generalization should not be made that mammals and birds have evolved different mechanisms for generating B cells. The bone marrow and the lymphoepithelial tissues in the gut probably both contribute to B-cell ontogeny in all species, to a greater or lesser extent. In the mouse and the rat, the bone marrow seems to have the major role (see review by OSMOND 1975). In the sheep and the fowl, the marrow is less well developed, and the focus is towards the lymphoid aggregates in the gut.

There is no obvious reason why B-cell genesis in a sheep should be more like that in a chicken than that in a mouse. However, some recent reports on the immunoglobulin gene in the chicken raise an intriguing possibility. REYNAUD et al. (1985) have shown that there are 8–12 germ-line genes for the variable region of the immunoglobulin light chain in the chicken. In man and the mouse, this region is encoded by about 50 and 200 genes, respectively (see review by HONJO 1983). Furthermore, the rearranged gene in almost all bursal B cells that were examined comprised the same joining-, constant-, and variable-region gene segments (REYNAUD et al. 1985). Clearly, the likelihood of generating immunoglobulin diversity by gene rearrangement is much less in the chicken than in the mouse, and probably man. Perhaps the bursa and PPs contribute to immunoglobulin diversity by providing a microenvironment that enables the continual turnover of B cells, thus promoting somatic mutation. Speculation along these lines would receive added support if sheep also contained a relatively small number of immunoglobulin variable-region genes. Unfortunately no information is available on this topic.

Acknowledgements. I am grateful to Linda Reynolds, Dean Befus, and Warren Chin for their help in preparing the manuscript. Research is currently supported by grants from the Alberta Heritage Foundation for Medical Research, and the Medical Research Council of Canada.

References

Ackerman A, Knouff RA (1959) Lymphocytopoiesis in the bursa of Fabricius. Am J Anat 104:163–205

Al Salami M, Simpson-Morgan MW, Morris B (1985) Haemopoiesis and the development of immunological reactivity in the sheep foetus. In: Morris B, Miyasaka M (eds) Immunology of the sheep. Roche, Basel, pp 19–36

Brambell FWR (1970) The transmission of passive immunity from mother to young. North-Holland, Amsterdam

Butcher EC, Rouse RV, Coffmann RL, Nottenburg CN, Hardy RH, Weissman IL (1982) Surface phenotype of Peyer's patch germinal center cells: implications for the role of germinal centers in B cell differentiation. J Immunol 129:2698–2707

Carlens O (1928) Studien über das lymphatische Gewebe des Darmkanals bei einigen Haustieren, mit Berücksichtigung der embryonalen Entwicklung, der Mengenverhältnisse und der Altersinvolution dieses Gewebes im Dünndarm des Rindes. Z Anat Entwicklungsgesch 86:393–493

Cooper MD (1983) Development of mature B cells from stem cells. In: Melchers F, Potter M, Weigert M (eds) Mechanisms of B cell neoplasia. Roche, Basel, p 47

Cooper MD, Lawton AR (1973) The mammalian "bursa equivalent": does lymphoid differentiation along plasma cell lines begin in the gut-associated lymphoepithelial tissues (GALT) of mammals? Contemp Top Immunobiol 1:49–68

Cornes JS (1965) Number, size and distribution of Peyer's patches in the human small intestine. Gut 6:225–233

Craig SW, Cebra JJ (1971) Peyer's patches: an enriched source of precursors for IgA-producing immunocytes in the rabbit. J Exp Med 134:188–200

Doughri AM, Altera KP, Kainer RA (1972) Some developmental aspects of the bovine fetal gut. Zentralbl Veterinar Med [A] 19:417–434

Ewijk W van (1984) Immunohistology of lymphoid and non-lymphoid cells in the thymus in relation to T lymphocyte differentiation. Am J Anat 170:311–330

Fahey KJ, Morris B (1978) Humoral immune responses in foetal sheep. Immunology 35:651–661

Ferguson A, Parrott DMV (1972) The effect of antigen deprivation on thymus-independent lymphocytes in the small intestine of the mouse. Clin Exp Immunol 12:477–488

Friedberg SH, Weissman IL (1974) Lymphoid tissue architecture. Ontogeny of peripheral T and B cells in mice: evidence against Peyer's patches as the site of generation of B cells. J Immunol 113:1477–1492

Gerber H, Morris B, Trevella W (1986) Role of gut-associated lymphoid tissues in the generation of immunoglobulin-bearing lymphocytes in sheep. Aust J Exp Biol Med Sci 64:201–213

Granfors K, Martin C, Lassila O, Suvitaival R, Toivanen A, Toivanen P (1982) Immune capacity of the chicken bursectomized at 60 hr of incubation: production of the immunoglobulins and specific antibodies. Clin Immunol Immunopathol 23:459–469

Grossi CE, Lydyard PM, Cooper MD, Franzi AT, D'Anna F (1977) Medullary localization of extracellular immunoglobulin and aminopeptidase in lymphoepithelial follicles of the chicken bursa of Fabricius and rabbit appendix. Dev Comp Immunol 1:157–164

Gulliani GL, Chanana AD, Cronkite EP, Joel DD (1974) Studies on lymphocytes. XI. Differences in the lymphocytopoietic activity of peripheral lymphoid organs. Proc Soc Exp Biol Med 145:1268–1271

Hofmann-Fezer G, Rodt H, Eulitz M, Thierfelder S (1976) Immumohistochemical identification of T- and B-lymphocytes delineated by the unlabelled antibody enzyme method. I. Anatomical distribution of θ-positive and Ig-positive cells in the lymphoid organs of mice. J Immunol Methods 13:261–270

Honjo T (1983) Immunoglobulin genes. Annu Rev Immunol 1:499–528

Huang HV, Dreyer WJ (1978) Bursectomy in ovo blocks the generation of immunoglobulin diversity. J Immunol 121:1738–1747

Husband AJ, Gowans JL (1978) The origin and antigen-dependent distribution of IgA-containing cells in the intestine. J Exp Med 148:1146–1160

Jalkanen S, Granfors K, Jalkanen M, Toivanen P (1983) Immune capacity of the chicken bursectomized at 60 hr of incubation: surface immunoglobulin and B-L (Ia-like) antigen-bearing cells. J Immunol 130:2038–2041

Jalkanen S, Jalkanen M, Granfors K, Toivanen P (1984) Defect in the generation of light chain diversity in bursectomized chickens. Nature 311:69–71

Jerne NK (1971) The somatic generation of immune recognition. Eur J Immunol 1:1–9

Kaminski M (1976) Transplantation of Peyer's patches under the kidney capsule in syngeneic mice. I. Morphological studies. Arch Immunol Ther Exp 24:759–768

Larsen HJ, Landsverk T (1986) Distribution of T and B lymphocytes in jejunal and ileal Peyer's patches of lambs. Res Vet Sci, 40:105–111

Lydyard PM, Grossi CE, Cooper MD (1976) Ontogeny of B cells in the chicken. I. Sequential development of clonal diversity in the bursa. J Exp Med 144:79–97

Mayrhofer G (1984) Physiology of the intestinal immune system. In: Newby TJ, Stockes CR (eds) Local immune responses of the gut. CRC, Boca Raton, pp 1–96

Mayrhofer GM, Pugh CW, Barclay AN (1983) The distribution, ontogeny and origin in the rat of Ia-positive cells with dendritic morphology and of Ia antigen in epithelia, with special reference to the intestine. Eur J Immunol 13:112–122

McElroy PJ, Wilcox N, Catty D (1981) Early precursors of B lymphocytes. I. Rabbit/mouse species differences in the physical properties and surface phenotype of pre-B cells, and in the maturation sequence of early B cells. Eur J Immunol 11:76–85

Melchers F (1979) Three waves of B-lymphocyte development during embryonic development of the mouse. In: Le Douarin N (ed) Cell lineage, stem cells and cell determination. Elsevier North-Holland, Oxford, pp 281–288

Meuwissen HJ, Kaplan GT, Perey DY, Good RA (1969) Role of rabbit gut-associated lymphoid tissue in cell replication. The follicular cortex as primary germinative site. Proc Soc Exp Biol Med 130:300–304

Miyasaka M, Dudler L, Bordmann G, Leiserson WM, Gerber H, Reynolds J, Trnka Z (1984) Differentiation of B lymphocytes in sheep. I. Phenotypic analysis of ileal Peyer's patch cells and the demonstration of a precursor population for sIg^+ cells in the ileal Peyer' patch. Immunology 53:515–523

Moore MAS, Owen JJT (1967) Experimental studies on the development of the thymus. J Exp Med 126:715–726

Opstelten D, Stikker R, Deenen GJ, Nieuwenhuis P (1982) Germinal centers and the B cell system. VII. Complement receptors, antigen receptors, immunoglobulin and alkaline phosphatase in germinal receptors on the rabbit appendix and popliteal lymph node. Cell Tissue Res 224:505–516

Osmond DG (1975) Formation and maturation of bone marrow lymphocytes. J Reticuloendothel Soc 17:97–112

Osmond DG, Nossal GJV (1974) Differentiation of lymphocytes in mouse bone marrow. II. Kinetics of maturation and renewal of antiglobulin-binding cells studied by double labelling. Cell Immunol 13:132–145

Owen JJT, Cooper MD, Raff MC (1974) *In vitro* generation of B lymphocytes in mouse foetal liver, a mammalian bursa equivalent. Nature 249:361–363

Pabst R, Reynolds JD (1986) Evidence of extensive lymphocyte death in sheep Peyer's patches. II. The number and fate of newly-formed lymphocytes that emigrate from Peyer's patches. J Immunol 136:2011–2017

Perey DYE, Good RA (1968) Experimental arrest and induction of lymphoid development in intestinal lymphoepithelial tissue of rabbits. Lab Invest 18:15–26

Pollard M, Sharon N (1970) Responses of Peyer's patches in germfree mice. Infect Immun 2:96–100

Reynaud C-A, Anquez V, Dahan A, Weill J-C (1985) A single rearrangement event generates most of the chicken immunoglobulin light chain diversity. Cell 40:283–291
Reynolds JD (1980) Gut-associated lymphoid tissues in lambs before and after birth. Monogr Allergy 16:187–202
Reynolds JD (1981) Peyer's patches and the ontogeny of B cells. In: Steinberg CM, Lefkovits I (eds) The immune system, vol 1. Karger, Basel, pp 95–101
Reynolds JD (1985) Evidence of differences between Peyer's patches and germinal centers. Adv Exp Med Biol 186:111–118
Reynolds JD (1986) Evidence of extensive lymphocyte death in sheep Peyer's patches. I. A comparison of lymphocyte production and export. J Immunol 136:2005–2010
Reynolds JD, Morris B (1983) The evolution and involution of Peyer's patches in fetal and postnatal sheep. Eur J Immunol 13:627–635
Reynolds JD, Morris B (1984) The effect of antigen on the development of Peyer's patches in sheep. Eur J Immunol 14:1–6
Reynolds JD, Pabst R (1984) The emigration of lymphocytes from Peyer's patches in sheep. Eur J Immunol 14:7–13
Reynolds JD, Cahill RNP, Trnka Z (1981) Peyer's patches as a bursal equivalent: a new look at some old arguments. In: Solomon JD (ed) Aspects of comparative and developmental immunology. Pergamon, Oxford, pp 265–272
Rose ML, Malchiodi F (1981) Binding of peanut lectin to thymic cortex and germinal centers of lymphoid tissues. Immunology 42:583–591
Scollay RG, Butcher EC, Weissman IL (1980) Thymus cell migration. Quantitative aspects of the cellular traffic from the thymus to the periphery in mice. Eur J Immunol 10:210–218
Silverstein AM, Uhr JW, Kraner KL, Lukes RJ (1963) Fetal response to antigenic stimulus. II. Antibody production by the fetal lamb. J Exp Med 117:799–812
Strober W, Jocobs D (1985) Cellular differentiation, migration, and function in the mucosal immune system. In: Gallin JI, Fauci AS (eds) Mucosal immunity. Raven, New York, pp 1–30
Wabl M, Burrows PD, von Gabain A, Steinberg C (1985) Hypermutation at the immunoglobulin heavy chain locus in a pre-B-cell line. Proc Natl Acad Sci USA 82:479–482
Waksman BH, Ozer H (1976) Specialized amplification elements in the immune system. Prog Allergy 21:1–113

B-Cell Commitment and Diversification in the Bursa of Fabricius

J.R.L. PINK[1,2] and O. LASSILA[1]

1 Introduction

The aim of this brief review is to draw attention to a possible relationship between three unusual features of the chicken immune system. First, the bird's B cells develop in a specialized organ with a follicular structure, the bursa of Fabricius. Secondly, the B-cell lineage, once it has been formed, is self-maintaining; that is, new B cells are not formed from immunoglobulin-negative (Ig^-) precursors at any time after the bird hatches. Finally, the chicken seems to use a very small number of Ig variable-region genes to generate a large antibody repertoire. We will discuss in detail only the first two of these points, as chicken Ig genes are the subject of another chapter in this volume (WEILL et al.). Our proposal, that there is a relationship between the number of Ig germ-line variable-region genes and the way in which B cells are generated in a given species, is based on recent articles dealing with chicken B-cell development and Ig gene structure (PINK et al. 1985a; WEILL et al. 1986, this volume).

2 B-Cell Development in the Bursa

The bursal rudiment develops as an outgrowth of the embryonic hindgut. From about day 8 to day 15 of embryonic incubation, lymphoid precursor cells migrate from extrabursal mesenchyme into the developing organ, and there proliferate in bursal follicles. These develop as invaginations of the folded epithelium, and are first visible as swellings in the epithelial layer on about day 12 of incuba-

[1] Basel Institute for Immunology, Grenzacherstr. 487, 4058 Basel, Switzerland
[2] Present address: Central Research Unit 68/21, Hoffmann-La Roche & Co., 4002 Basel, Switzerland

Current Topics in Microbiology and Immunology, Vol. 135

Table 1. Fraction of follicles stained with anti-Bu-1a antibody in bursae of chicken embryos parabiosed at different ages

Embryo Bu-1a type (+ or −)	Age of embryo at parabiosis (days)	No. of pairs parabiosed	Percentage of follicles in each embryo stained with anti-Bu-1a					
+	8	4	100	100	92	65		
−	8		100	94	73	31		
+	9	6	99	98	98	97	97	92
−	9		99	96	94	93	84	89
+	10	3	94	93	83			
−	10		75	59	44			
+	11	4	100	100	98	74		
−	11		53	20	10	36		
+	9	2	96	69				
−	10		54	56				
+	10	3	100	100	92			
−	9		100	98	70			

H.B21 embryos of opposite Bu-1a type, parabiosed at the stated age, were examined at or just before hatching. Frozen bursal sections were stained with anti-Bu-1a monoclonal antibody and fluorescent goat anti-mouse Ig. The average number N of precursor cells populating a follicle can be estimated from the equation $1/2^{N} = F^{+}/(1\text{-}F^{-} + F^{+})$, where F^{+} and F^{-} are the fractions of "negative" follicles (not containing Bu-1a^{+} cells) in age-matched Bu-1a^{+} and Bu-1a^{-} embryos respectively. Data from Pink et al. (1985b) and unpublished observations

tion (Houssaint et al. 1976; Lassila et al. 1979). In all, about 10^4 follicles are formed during the colonization of the bursa by lymphoid stem cells, and this number remains constant as the bird matures (Olah and Glick 1978).

The number of lymphoid stem cells populating a follicle is small (less than 10) (Pink et al. 1985b). This was estimated by examining sections of bursae from parabiosed chicken embryos differing at the *Bu-1* locus, which codes for a B-cell alloantigen. The numbers of follicles reacting with a monoclonal antibody against one allelic *Bu-1* product, the Bu-1a antigen, in bursae of parabiosed Bu-1a^{+} and Bu-1a^{-} embryos are shown in Table 1. From these numbers it was estimated (see Table 1) that the average number of cells populating a follicle is three to four; however, this number should be considered a rough estimate only, in view of the various assumptions made in its derivation (Pink et al. 1985b; Pink 1986).

Little is known about the phenotype of the lymphoid stem cells which populate the normal bursa. They probably do not express Ig (Szenberg 1976; Ewert and Cooper 1978) and may be multipotential cells with unrearranged Ig genes, as suggested by the finding that at least some cells from an 11-day quail bursa can migrate into a chicken thymus in an organ-culture system (Jotereau et al. 1980). However, other experiments suggest that the precursor cells are either committed to the B-cell lineage, or become so shortly after entering the bursa (see Sect. 3). In addition, studies of bursal cells from 13- and 14-day embryos and of bone marrow cells from embryos of 14 days or more indicate that, at least during late embryonic development, there exist separate lineages of cells capable of repopulating either the bursa or the thymus (Weber and Alexander 1978; Weber and Foglia 1980; Brand et al. 1983).

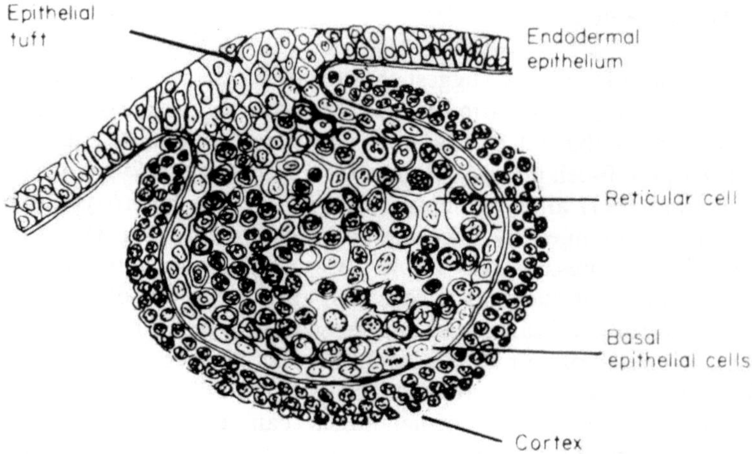

Fig. 1. Structure of a mature bursal follicle (from WHITE 1981). The bursal medulla (*center*) contains epithelial reticular cells which are loosely mixed with lymphocytes and macrophages. The medulla is bounded by a spherical bag of epithelial cells and a basement membrane, which separates the medulla from the cortex containing densely packed lymphocytes

It is clear that lymphoid cell proliferation in separate follicles proceeds independently, with very little traffic of cells from one follicle to another, since neighboring follicles in chimeric birds often contain cell populations of distinct origins (PINK et al. 1985b). The proliferation is very rapid, with a population-doubling time of about 10 h in 14 to 16-day chicken embryos (LYDYARD et al. 1976). Ig-bearing (Ig^+) cells are first visible within the bursa at about day 12 of incubation, and by day 18 about 85% of cells in a bursal cell suspension are Ig^+ (LYDYARD et al. 1976; SZENBERG 1976; EWERT and COOPER 1978). From about this time on, B cells are exported from the organ (WOODS and LINNA 1965; GLICK 1977; IVANYI 1981), the fraction exported being about 1%–5% of the cells produced, with the rest of the newly produced cells presumably dying in situ (O. LASSILA, unpublished observations). The bursa increases to a maximum weight, 5–7 weeks after hatching, of about 2–3 g in White Leghorn birds (GLICK 1956), at which stage it contains about 3×10^9 lymphocytes (3×10^5 per follicle), of which at least 90% carry cell surface IgM (IgM^+) (LYDYARD et al. 1976). A small percentage (<1%) carry IgG and IgA; as these appear early in ontogeny (CHEN 1978), they may be formed in the bursa by intrabursal switching from IgM^+ precursors.

The structure of a mature follicle is shown in Fig. 1. The lymphoid cells which develop in the embryonic follicle can traverse the basal epithelial cell layer, so that by the time of hatching the follicle consists of a cortex and a medulla separated by the epithelium; the individual follicles are separated by a connective-tissue boundary (OLAH et al. 1975). Thus, cortical and medullary cells in the same follicle are derived from the same precursors, but otherwise the relationship between cells in the two compartments is unclear (COOPER et al. 1972).

3 B-Cell Self-Renewal

Three lines of evidence suggest that avian B cells are self-renewing, in the sense that there is no generation of B cells from Ig^- precursors in adult birds (we do not imply that all B cells in an adult bird contribute equally to the maintenance of the B-cell lineage). RATCLIFFE and IVANYI (1979) showed that long-term IgM (and IgG) allotype suppression in allotype-heterozygous chickens could readily be obtained by injecting embryos with appropriate anti-IgM allotype sera. The suppression was stable for at least 1 year and was probably not due to the action of suppressor cells, but to a depletion of bursal and peripheral B-cell precursors bearing the suppressed allotype. These experiments suggested that by day 18 of embryonic incubation B cells were no longer being generated from Ig^- precursors (RATCLIFFE and IVANYI 1979, 1981; RATCLIFFE et al. 1986).

B-cell precursors in normal bursae can repopulate B-cell-deficient bursae and restore the peripheral B-cell function of cyclophosphamide-treated or irradiated recipients. These bursal progenitor cells are derived from the lymphoid stem cells which enter the bursa during embryonic development, and can be detected in the bursa from day 13 of incubation until about 4–5 weeks after hatching (TOIVANEN and TOIVANEN 1973; WEBER and ALEXANDER 1978). A useful assay system for bursal progenitor cells is their ability to reconstitute bursal weight in cyclophosphamide-treated recipients. Incubation of bursal cells from normal 4-day-old chickens with anti-Ig sera inhibits their ability to reconstitute in this assay, suggesting that the bursal progenitor cells carry cell surface Ig. Also, when bursal cells from IgM-allotype-heterozygous donors were treated with anti-IgM allotype serum before reconstitution, the appearance of the corresponding allotype-bearing cells in the host's bursa was prevented (Table 2), and the relevant serum allotype was suppressed (PINK et al. 1985a). These observations, and the fact that the reconstituted follicles never contain more than a few percent of surface IgG- or IgA-bearing cells (O. LASSILA, unpublished observations) suggest that all bursal progenitor cells in neonatal chicks are surface IgM^+.

Finally, a study of IgM allotype expression in single follicles of normal, 4–10-week-old, allotype-heterozygous birds showed that in many follicles cells of one or other allotype predominated (Table 2), suggesting that new Ig gene rearrangements (which would presumably produce cells carrying the two allotypes in roughly equal proportions) do not take place in the bursa to any appreciable extent after hatching (RATCLIFFE et al. 1986). Very similar results were obtained when gene rearrangements were examined directly in individual follicles, using DNA probes detecting either light-chain or heavy-chain gene rearrangements (WEILL et al. 1986). From the number of follicles which appeared not to be mixed but to contain cells with one allotype or one pattern of gene rearrangement, it seems likely that a very small number of independent rearrangements (less than 10) contributes to the Ig diversity in the follicle. The number n of independent rearrangements can be estimated from the formula $N_A{}^{1/n} + N_B{}^{1/n} = 1$, where N_A and N_B are the proportions of follicles which appear homogeneous and are of type A and B respectively (A and B can be, for example, two allotypes or two allelic variants of a rearranged light- or heavy-chain gene).

Table 2. Immunoglobulin allotypes in individual bursal follicles of normal or cyclophosphamide-treated, bursa cell-reconstituted chickens

No. and type of bursae examined	Total no. of follicles examined	No. of follicles of each type observed			Mean percentage of M-1 a/M-1 b cells in the mixed follicles
		M1 a	M1 b	Mixed	
5 normal (*M-1*^{a} × *M-1*^{b}) F_1 bursae	71	5	21	45	41/51
3 bursae reconstituted with normal serum-treated F_1 cells	48	19	21	8	23/46
1 bursa reconstituted with anti-M-1 a-treated F_1 cells	13	1	9	3	12/75
2 bursae reconstituted with anti-M-1 b-treated F_1 cells	24	21	3	0	–

The bursae examined were from normal (H.B14B × H.B14C)F_1 birds (allotype *M-1*^{a} × *M-1*^{b}), or from H.B14A (allotype *M-1*^{a}) birds which had been treated with cyclophosphamide at hatching and then reconstituted with (H.B14B × H.B14C)F_1 bursal cells (allotype *M-1*^{a} × *M-1*^{b}). The reconstituting cells were incubated with 10% normal chicken serum, or anti-M-1 a or anti-M-1 b serum, immediately before being injected. Individual follicles were isolated 10 weeks later and were used to prepare single-cell suspensions for immunofluorescent staining. Similar suspensions were made from bursae of normal 4- to 10-week-old birds. Aliquots of each suspension were stained with affinity-purified anti-M-1 a and anti-M-1 b antibodies using a triple-layer, indirect immunofluorescence assay. The percentages of cells stained with each reagent were determined by counting at least 100 cells for each follicle, and the values obtained with the second and third reagents alone were taken as background. Follicles in which <7% of cells stained with anti-M-1 b or anti-M-1 a, after correction by subtraction of the background staining for that follicle, were classed as "M-1 a" or "M-1 b" follicles respectively; the remaining follicles were classed as mixed. The mean percentages of M-1 a$^+$ and M-1 b$^+$ cells in cell suspensions made from large fragments of the normal (M-1 a × M-1 b) bursae were 37% and 60% respectively. Data from PINK et al. (1985 a), RATCLIFFE et al. (1986) and unpublished results

The value of *n* was estimated as about 2–3 from two sets of experiments; this number is close to the number of cells populating the follicle (see Sect. 2), suggesting that gene rearrangement occurs at the time of, or shortly after, the entry of the precursors into the follicle. In this calculation it is assumed that no selection for particular cell types takes place during their proliferation in the follicle; since this possibility cannot be excluded (and is even *a priori* quite likely), the numbers given above should be taken as rough estimates only (RATCLIFFE et al. 1986; WEILL et al. 1986; PINK 1986).

4 The Bursa as a Mutant-Generating Organ

Evidence that mutants are generated in the bursa comes from two sources: the effect of bursectomy on antibody diversity, and the demonstration of Ig variability in individual follicles populated by single clones of B cells. Embryos surviving the removal of the tail region after 60 h of incubation are incapable of making specific antibodies to a variety of antigens (JALKANEN et al. 1984), and, as measured by isoelectric focusing (IEF), have reduced Ig light-chain diversity in their serum IgG, although they may have almost normal levels of IgA and IgM. These Igs may be derived from a very few clones which have developed, but not undergone mutation, in the absence of the bursa. Hormonal bursectomy of 11-day-old embryos also results in a reduction of Ig light-chain diversity as measured by IEF (HUANG and DREYER 1978).

When a neonatally cyclophosphamide-treated chick is reconstituted with normal bursal cells, the majority of its 10^4 follicles are repopulated by single Ig^+ cells (PINK et al. 1985a, b). Nevertheless, the reconstituted bird responds well to various antigens (TOIVANEN and TOIVANEN 1973; O. VAINIO, unpublished observations), and the Ig light chains synthesized by the 3×10^5 cells present in a single follicle 6 weeks after reconstitution are almost as diverse (as estimated by two-dimensional gel electrophoresis) as total bursal Ig light chains. Also, different heavy-chain variable-region (V_H) idiotypic determinants are found on different cells in single reconstituted follicles (LASSILA et al. 1986; PINK et al. 1986). In fact, idiotype-bearing or antigen-binding cells are present in small numbers (e.g., $\leq 1\%$ of cells in the follicles) even in many normal follicles (LYDYARD et al. 1976; O. LASSILA et al., unpublished observations); for one antigen (sheep red blood cells) the proportion of antigen-binding cells was shown to increase during ontogeny (LYDYARD et al. 1976). These observations, together with evidence that B-cell proliferation in the bursa is antigen independent (LYDYARD et al. 1976; LUPETTI and DOLFI 1980), suggest that the role of the bursa is to act as a mutant-generating organ.

5 Concluding Comments

The features of the chicken immune system discussed above – the early development of B cells in a compartmentalized, mutant-generating organ, and their subsequent self-renewal – may be related to the small number of Ig variable-region genes used by the bird in the following way: one might predict that generation of antibody diversity from a few variable-region genes would require extensive antigen-independent B-cell proliferation, accompanied by somatic mutation early in ontogeny; that the proliferation would take place in a compartmentalized organ, in order to avoid competition between expanding clones; and that once generated, B cells would be self-renewing, so that mutants are not lost, but accumulate with time.

Our argument would be strengthened if it could be shown that there was a correlation between the number of variable-region genes and the extent of B-cell self-renewal in different species. It is clear that the number of variable-region genes can vary widely in different species. Laboratory mice are thought to have about 200 κ-chain variable-region (V_κ) genes and a similar number of V_H genes. The numbers of genes in humans are lower, perhaps by an order of magnitude (HONJO 1983). In both mice and humans extensive variability can be generated simply by gene rearrangement; for example, the repertoire of mouse antibodies in the total absence of somatic mutation would be greater than 10^7 if all possible combinations of V_H and light-chain variable region (V_L), and of variable–joining regions (V-J) or variable–diverse–joining regions (V–D–J) were to be used (Tonegawa 1983). In these circumstances, a large Ig repertoire can be produced and maintained by continuous formation of B cells by fresh gene rearrangements, with little clonal proliferation and mutation of the newly generated cells. Support for the idea that antigen-independent somatic mutation in mice may be relatively less important than in chickens comes from consideration of κ-suppressed mice. Like chickens, these have to rely on a very small λ-chain variable-region (V_λ) germ-line gene repertoire, yet their primary antibody responses to various antigens are low (Weiss et al. 1984).

It is also clear that species differ in the way that B cells are generated. In mice, the primary site of B-cell formation is probably the fetal liver; later, B cells are produced in the bone marrow from cell surface Ig^- precursors, even in adult animals. We assume here that precursors with unrearranged Ig variable-region genes make an important contribution to this B-cell production, although this is by no means firmly established. In sheep, on the other hand, the Peyer's patches are a major site of B-cell development. The similarities between the bursa and sheep Peyer's patches, for example in follicular structure and high rate of B-lymphocyte production in fetal or embryonic life, are striking and are discussed further by Reynolds in this volume. These observations suggest that there are two contrasting ways of generating antibody variability, a "big-bang" model (favored in chickens and possibly sheep) in which B cells are formed during a short period of ontogenesis and thereafter accumulate somatic mutations in a small number of variable-region genes, and a "steady-state" model (favored by mice) in which variability is generated by the continual formation of B cells with fresh rearrangements of large numbers of Ig variable-region genes.

References

Brand A, Galton J, Gilmour DG (1983) Committed precursors of B and T lymphocytes in chick embryo bursa of Fabricius, thymus, and bone marrow. Eur J Immunol 13:449–453

Chen CLH (1978) Ontogeny of immunoglobulin isotype diversity expressed by chicken lymphocytes. Fed Proc 37:1395

Cooper MD, Lawton AR, Kincade PW (1972) A developmental approach to the biological basis for antibody diversity. Contemp Top Immunobiol 9:33–47

Ewert DL, Cooper MD (1978) Ia-like alloantigens in the chicken: serologic characterization and ontogeny of cellular expression. Immunogenetics 7:521–535

Glick B (1956) Normal growth of the bursa of Fabricius in chickens. Poultry Sci 35:843–851

Glick B (1977) The bursa of Fabricius and immunoglobulin synthesis. Int Rev Cytol 48:345–402

Honjo T (1983) Immunoglobulin genes. Annu Rev Immunol 1:499–528

Houssaint E, Belo M, Le Douarin NM (1976) Investigations on cell lineage and tissue interactions in the developing bursa of Fabricius through interspecific chimeras. Dev Biol 53:250–264

Huang HV, Dreyer WJ (1978) Bursectomy in ovo blocks the generation of immunoglobulin diversity. J Immunol 121:1738–1747

Ivanyi J (1981) Functions of the B lymphoid system in chickens. In: Rose ME, Payne LN, Freeman BM (eds) Avian immunology. British Poultry Science, Edinburgh, pp 63–101

Jalkanen S, Jalkanen M, Granfors K, Toivanen P (1984) Defect in the generation of light chain diversity in bursectomised chickens. Nature 311:69–71

Jotereau FV, Houssaint E, Le Douarin NM (1980) Lymphoid stem cell homing to the early thymic primordium of the avian embryo. Eur J Immunol 10:620–626
Lassila O, Eskola J, Toivanen P (1979) Prebursal stem cells in the intraembryonic mesenchyme at 7 days of incubation. J Immunol 123:2091–2094
Lassila O, Alanen A, Lefkovits I, Pink JRL, Cooper MD (1987) Immunoglobulin diversification in vivo within intrabursal clonal progeny of precursor B cells (to be published)
Lupetti M, Dolfi A (1980) A contribution to the study of lymphopoiesis in the bursa of Fabricius in Gallus Domesticus. Transplantation 29:67–71
Lydyard PM, Grossi CE, Cooper MD (1976) Ontogeny of B cells in the chicken. I. Sequential development of clonal diversity in the bursa. J Exp Med 144:79–97
Olah I, Glick B (1978) The number and size of the follicular epithelium and follicles in the bursa of Fabricius. Poultry Sci 57:1445–1450
Olah I, Röhlich P, Töro I (1975) Ultrastructure of lymphoid organs. An electron microscope atlas. Masson, Paris
Pink JRL (1986) Counting components of the chicken's B cell system. Immunol Rev 91:115–128
Pink JRL, Ratcliffe MJH, Vainio O (1985a) Immunoglobulin-bearing stem cells for clones of B (bursa-derived) lymphocytes. Eur J Immunol 15:617–620
Pink JRL, Vainio O, Rijnbeek A-M (1985b) Clones of B lymphocytes in individual follicles of the bursa of Fabricius. Eur J Immunol 15:83–87
Pink JRL, Lassila O, Vainio O (1987) B lymphocytes and their self-renewal In: Toivanen A, Toivanen P (eds) Avian immunology. Basis and practice. CRC, Boca Raton (to be published)
Ratcliffe MJH, Ivanyi J (1979) Allotype suppression in the chicken. I. Generation of chronic suppression in heterozygous but not in homozygous chickens. Eur J Immunol 9:847–852
Ratcliffe MJH, Ivanyi J (1981) Allotype suppression in the chicken. IV. Deletion of B cells and lack of suppressor cells during chronic suppression. Eur J Immunol 11:306–310
Ratcliffe MJH, Lassila O, Pink JRL, Vainio O (1986) Avian B cell precursors. Immunoglobulin expression is an early, possibly bursa-independent event. Eur J Immunol 16:129–133
Szenberg A (1976) Ontogenesis of the immune system in birds. In: Marchalonis JJ (ed) Comparative immunology. Blackwell, Oxford, pp 419–431
Toivanen P, Toivanen A (1973) Bursal and postbursal stem cells in chicken. Functional characteristics. Eur J Immunol 3:585–595
Tonegawa S (1983) Somatic generation of antibody diversity. Nature 302:575–581
Weber WT, Alexander JE (1978) The potential of bursa-immigrated hematopoietic precursor cells to differentiate to functional B and T cells. J Immunol 121:653–658
Weber WT, Foglia LM (1980) Evidence for the presence of precursor B cells in normal and in hormonally bursectomized chick embryos. Cell Immunol 52:84–94
Weill JC, Reynaud CA, Lassila O, Pink JRL (1986) Rearrangement of chicken immunoglobulin genes is not an ongoing process in the embryonic bursa of Fabricius. Proc Natl Acad Sci USA 83:3336–3340
Weiss S, Lehmann K, Raschke WC, Cohn M (1984) Mice completely suppressed for the expression of immunoglobulin κ light chain. Proc Natl Acad Sci USA 81:277–215
White RG (1981) The structural organisation of avian lymphoid tissues. In: Rose ME, Payne LN, Freeman BM (eds) Avian immunology. British Poultry Science, Edinburgh, pp 21–43
Woods R, Linna J (1965) Transport of cells from the bursa of Fabricius to the spleen and the thymus. Acta Pathol Microbiol Scand 64:470–476

Differentiation of Murine B-Cell Progenitors in Agar Culture: Determination of the Developmental Potential of Clonable Pre-B Cells

H. Sauter and C.J. Paige

1 Introduction

The differentiation of stem cells into mature B lymphocytes comprises a sequence of developmental steps which leads to cells expressing unique immunoglobulin (Ig) molecules. Much has been learned about intermediate stages in this pathway through the application of a variety of experimental approaches to both transformed cell lines and populations of normal B-cell precursors.

Of central importance has been the identification of the genetic rearrangements by which those gene segments which encode the variable and constant regions of the Ig molecule are combined (for review see Tonegawa 1983). These analyses have primarily relied upon homogeneous cell lines which are representative of different developmental stages. It appears from these studies that the rearrangement and expression of Ig genes is a developmentally ordered process (for review see Alt et al. 1984). For example, one of the earliest cells in the B-lymphoid pathway produces cytoplasmic μ heavy chains but has not yet rearranged its light chain genes (Burrows et al. 1979; Maki et al. 1980; Perry et al. 1981; Alt et al. 1981). This finding has been confirmed for normal B cells using enriched populations of B-cell precursors (Coffman and Weissman 1983). The expression of complete surface Ig further requires rearrangement and expression of light chain genes. The analysis of κ-producing myelomas, hybridomas, and splenic B cells suggests further that the rearrangement usually occurs first at the κ locus and, in case of abberrant κ rearrangements, subsequently at the λ locus (Alt et al. 1981; Coleclough et al. 1981). However, exceptions to that rule have been observed (Coleclough et al. 1981). Although consensus has been achieved regarding the pattern of Ig gene rearrangements, the elements

Basel Institute for Immunology, Grenzacherstrasse 487, 4058 Basel, Switzerland

Current Topics in Microbiology and Immunology, Vol. 135

which control the regulation of the rearrangement and subsequent expression of Ig genes remain poorly understood. Some of these questions are addressed in the following chapters. In addition, it is uncertain whether cells defined by these rearrangements represent stable intermediates or rather progress rapidly through these steps unless trapped by immortalization. Our understanding of the emergence, frequency, absolute number, and growth requirements of the various intermediates is also incomplete.

Many of these questions have been addressed in functional assay systems, which monitor the emergence of B cells from precursor cells either by in vivo reconstitution or by in vitro culture (for review see KINCADE 1981). Cell transfer experiments identified 7- to 9-day yolk sac as the earliest site with lymphoid potential in murine development (TYAN and HERZENBERG 1968; PAIGE et al. 1979). Developmental differences between precursor populations were revealed in reconstitution experiments comparing yolk sac cells and fetal liver cells. These experiments suggested that yolk sac cells might be a less differentiated population of precursors (PAIGE et al. 1979). Further support for this interpretation came from experiments using in vitro culture techniques, which showed that the lymphoid potential of yolk sac, in contrast to that of fetal liver, could not be demonstrated in these assay systems (OWEN et al. 1975; PHILLIPS and MELCHERS 1976; MELCHERS and ABRAMCZUK 1980). A further example for the usefulness of cell transfer assays comes from experiments in which the relationship of B cells to other hematopoietic lineages has been examined. For example, it could be shown that stem cells giving rise to colonies in the spleen of an irradiated host (CFU-S) are most likely not progenitors of the B-lineage cells (PAIGE et al. 1981). Irradiation/reconstitution studies were also essential in demonstrating the existence of multipotential stem cells in bone marrow. Initially using chromosomally marked cells, and more recently cells marked by retrovirus insertion, stem cells with the capacity to generate both myeloid and lymphoid progeny have been demonstrated (WU et al. 1968; ABRAMSON et al. 1977; DICK et al. 1985; KELLER et al. 1985).

While it has been possible to illuminate the functional potential of precursor cells by in vivo reconstitution, more controlled conditions have been obtained in liquid culture with in vitro systems using either organ fragments or single-cell suspensions. These techniques have been used to examine accessory cell function in B-cell development and to begin to characterize the requirements needed for the transition from a precursor to a mature B cell.

The development of functional B cells from fetal liver has been observed in several experimental systems in which support for B-cell development was provided by diverse sources of cells, including fetal liver cells themselves (OWEN et al. 1975), splenocytes (TEALE and MANDEL 1980), thymocytes (MELCHERS 1977a, b, 1979), and bone marrow cells (KINCADE et al. 1981a). However, the essential cells in those populations remain poorly characterized and thus it is not yet clear whether they are relevant to normal development. It is even uncertain whether accessory cell function requires cell-to-cell contact or relies on soluble growth and/or differentiation factors. Several molecules have been identified which might have regulatory effects on B-cell precursors, including interleukin-1 (HOFFMANN 1980; GIRI et al. 1984), interleukin-3 (PALACIOS et al.

1984) or γ-interferon (GIRI et al. 1984). In many cases the effect of these regulatory molecules has been demonstrated by using homogeneous cell lines. Evidence has also been presented showing an effect on normal cell populations, but usually under conditions which fail to distinguish between direct and indirect effects. In at least one asssay system the support by growth factors such as interleukin-3, granulocyte and macrophage colony-stimulating factor (GM-CSF), and colony-stimulating factor 1 (CSF-1) was shown to be indirect (PAIGE et al. 1985 a).

One promising approach for distinguishing possible intermediate developmental stages in the B-cell pathway has been the utilization of antibodies against cell surface proteins, which distinguish subpopulations of B cells and might be expressed at different stages during B-cell differentiation. A surface marker of particular interest has been the B220 form of the *Ly-5* locus (KINCADE et al. 1981 b; COFFMAN and WEISSMAN 1981). This antigen is present on B cells and also early B-cell precursors (SUGIYAMA et al. 1982; COFFMAN and WEISSMAN 1983). Another useful marker is the Lyb-2 antigen which is expressed on B cells and some B-cell precursors and seems to be restricted to the B lineage (SATO and BOYSE 1976; KINCADE et al. 1981 b). Other monoclonal antibodies (MAbs) directed against antigens not exclusively expressed on B-lineage cells are also helpful in defining subpopulations of cells, for example the MAbs AA4 (MCKEARN et al. 1984) and 19B5 (SHINEFELD et al. 1980; PAIGE et al. 1981).

Although considerable progress has been made in our understanding of B-cell development through the use of these different experimental systems, many relevant aspects are still poorly understood. It has been particularly difficult to apply the knowledge derived from the molecular analysis of long-term cell lines directly to normal B-cell precursors, which are usually rare in natural cell populations. Consequently, the genetic state of the precursors being detected in functional assay systems remains obscure. Furthermore, the growth requirements for development and expansion along the B-cell pathway are still poorly defined.

To be able to examine these questions, an assay system is required which (1) is clonal and quantitative for B-cell progenitors, (2) allows B-cell precursors to differentiate into mature B cells whose properties can be studied, (3) can determine the developmental stage of the B-cell precursor, and (4) allows us to identify the requirements for the transition along the B-cell pathway. We have described such a clonal system for B-cell precursors which differentiate into colonies containing Ig-secreting cells (PAIGE 1983; PAIGE et al. 1984, 1985 a, b).

In this article we discuss the general characteristics of clonable pre-B cells regarding frequency and surface characteristics. Further, we describe a more recent modification of the assay which allows us to more carefully assess the developmental potential of these cells.

2 The Clonable Pre-B Cell Assay

The clonable pre-B cell assay is a modification of previously described methods for the growth of B lymphocytes in agar (METCALF et al. 1975; KINCADE et al.

1976) and for detecting Ig-secreting B-cell colonies (PAIGE and SKARVALL 1982). The cells which initiate colony formation in the clonable pre-B cell assay can be distinguished from clonable B cells by several criteria, including (1) presence in 12- to 16-day fetal liver, (2) delayed onset of Ig secretion, (3) delayed reactivity to lipopolysaccharide (LPS), (4) adherent cell dependence, (5) lack of sIg, and (6) rapid sedimentation rates. These precursors give rise within approximately 1 week to colonies containing up to 2000 cells. Some cells within these colonies are secreting IgM, which can be detected either by a Protein-A plaque assay (PAIGE and SKARVALL 1982) or by an enzyme-linked immunoassay using nitrocellulose filters to which the Ig has bound (SAUTER and PAIGE 1986, to be published). The principal advantage of the nitrocellulose method is that it permits replicate blots of single colonies and therefore facilitates the detection of multiple Ig isotypes within these clones. Experimental details for these procedures have been published elsewhere (PAIGE et al. 1985b, 1986; SAUTER and PAIGE 1986, to be published).

3 The Clonable Pre-B Cell Assay Is Limiting for the B-Cell Precursor

In order to determine frequencies and absolute numbers of B-cell precursors in a heterogeneous cell population, and to distinguish between direct and indirect influences of growth stimulation, an assay system should be limiting for the precursor cell itself. Several lines of evidence indicate that the clonable pre-B cell assay fulfills this requirement. For cell concentrations up to 2×10^5 cells per culture, the log–log relationship between input cell number and number of secreting colonies subsequently detected has a slope of one, suggesting that a single component is limiting in the system. This finding is consistent with the interpretation that a colony is derived from a single cell. An alternative explanation would be that an accessory cell is limiting and stimulates several precursor cells to form colonies. This seems to be unlikely, since precursor-enriched cell populations selected for the expression of B-lineage-specific antigens such as B220 or Lyb-2 also respond with a slope of one (PAIGE et al. 1985a). Under these conditions, up to 80% of the colonies generated secrete Ig and almost all of them show the typical morphology of B-cell colonies. Further proof that an accessory cell is not limiting in the assay comes from mixing experiments utilizing congenic mouse strains. The rat-anti-mouse μ MAb 33-24-12 does not recognize IgM of C.AL 20 origin, but does recognize IgM of most other strains, including BALB/c (GRÜTZMANN 1981). Fetal liver cells from these two strains were mixed at different ratios to allow random interaction between accessory cells and precursor cells. If the precursor cell is the limiting cell, the addition of C.AL 20 cells to BALB/c cells should not affect the number of 33-24-12-specific plaques detected, but should result in an increase in the number of Protein-A plaques found. In contrast, if the accessory cell is limiting, then the number of 33-24-12-specific plaques should also be increased. The results show that addition of C.AL 20 cells to BALB/c cells caused an increase only in the number of Protein-A plaques, providing further evidence that the clonable pre-B cell assay is limiting for the precursor cell itself.

3.1 Frequency and Absolute Number of Clonable Pre-B Cells During Ontogeny

The assay therefore allows us to determine the frequencies and absolute number of clonable pre-B cells during development. The estimates for the frequencies of clonable pre-B cells in fetal liver are: 1 in 300000 on day 12, 1 in 84000 on day 13, 1 in 19000 on day 14, 1 in 8000 on day 15, and 1 in 6000 cells on day 16 of gestation (PAIGE et al. 1984). These frequencies have been determined using the Protein-A plaque assay. Comparable results are obtained with the nitrocellulose transfer assay (SAUTER and PAIGE 1986, to be published). The frequencies, together with the total number of cells recovered, allow us to estimate the absolute number of clonable pre-B cells in fetal liver-during ontogeny. These estimates are: approximately 10 on day 12, 100 on day 13, 1000 on day 14, 4500 on day 15, and 5000 on day 16. Thus, the absolute number of clonable pre-B cells increases five- to tenfold daily from day 12 to day 15, reaching maximum values on day 16 in ontogeny.

4 Surface Characteristics of Clonable Pre-B Cells

The cell surface characteristics of clonable pre-B cells have been examined by exposing fetal liver cells to petri dishes coated with antibodies specific for cell surface molecules. Cells expressing the particular antigen bind to the dishes and can be separated from cells which fail to express it (WYSOCKI and SATO 1978). It should be noted that the "panning" technique depends on several variables, such as the number of molecules present on the cell surface, the affinity of the antibodies employed, and possible influences of the antibodies on further differentiation steps. Nonetheless, our results are highly reproducible and are thus useful for distinguishing subsets of B-cell progenitors (PAIGE et al. 1985a).

Experiments with anti-μ-coated dishes have shown that no reduction in the number of secreting colonies was obtained with fetal liver cells from day 13 to day 16 of gestation. In contrast, almost all colony-forming B cells from spleen could be removed by this technique. The antibodies used to characterize clonable pre-B cells were 14.8 (anti-B220) (KINCADE et al. 1981b), AA4 (determinant unknown) (MCKEARN et al. 1984), and 10.1.D2 (anti-Lyb-2) (SUBBARAO and MOSIER 1982).

The experiments have shown that there is a temporal appearance of these markers during ontogeny – none of them could be detected on precursors from 12-day fetal liver, and only AA4 is expressed on day 13. Over 80% of the precursors from day 14 express AA4 and about 70% of them are 14.8-positive. Precursors from 15- and 16-day fetal liver show the same expression pattern for AA4 and B220 as 14-day fetal liver precursors, but in addition some of them (30%–40%) are positive for Lyb-2. Based on these studies at least four different subpopulations can be defined: $AA4^-$, 14.8^-, Lyb-2^-; $AA4^+$, 14.8^-, Lyb-2^-; $AA4^+$, 14.8^+, Lyb-2^- and $AA4^+$, 14.8^+, Lyb-2^+. It is not yet clear whether this indicates a parent–progeny relationship between these subsets or whether they are unrelated subpopulations of precursor cells.

5 Light Chain Commitment of Clonable Pre-B Cells

The initial characterization of clonable pre-B cells revealed properties which suggested that they were very early members of the B lineage. Therefore, it was of interest to determine their developmental stage more precisely and define their differentiation options. To study this question we utilized a nitrocellulose blotting technique (SAUTER and PAIGE 1986, to be published). In this method the top agar layer containing the Ig-secreting colonies was transferred to a glass slide, dehydrated, and overlaid with a pre-wetted nitrocellulose filter. Secreted Ig diffuses from the agar matrix to the nitrocellulose filter, where it can be visualized by an enzmye-linked immunoassay using isotype-specific MAbs. Since replicates can be taken from the same colony, this method allowed us to examine the question of whether different types of light chains are expressed within one colony. Commitment to a particular light chain at the initiation of culture will result in the generation of colonies of cells homogeneous regarding light chain isotype. In contrast, a clonable pre-B cell which is not yet committed to a particular light chain gene at the initiation of culture, may develop into a colony in which both κ and λ light chains are expressed.

In the adult mouse about 95% of all B cells express κ light chains and only 5% express λ light chains (MCGUIRE and VITETTA 1981; KESSLER et al. 1981; TAKEMORI and RAJEWSKY 1981). This has been attributed either to the higher probability of successful κ gene rearrangements, perhaps due to the greater number of V_κ genes (VALBUENA et al. 1978; TONEGAWA et al. 1978), or to a selection mechanism which favors κ-expressing B cells (HAUGHTON et al. 1975; NISHIKAWA et al. 1984). We found that colonies from mature clonable B cells also expressed a predominance of κ light chains (SAUTER and PAIGE 1986, to be published). In adult splenic B-cell derived colonies, about 3%–4% of the colonies secreted λ light chains, in good agreement with results from other experimental systems. Similarly, we found that λ expression in colonies derived from progenitors in 13- to 16-day fetal liver ranged between 5% and 9% of the μ-secreting colonies. These results strongly suggest that κ predominance is established early in development prior to selection by antigen.

We also observed that some of the precursor-derived colonies secreted both κ and λ light chains at the same time. From the 13th to the 15th day of gestation $\sim$5% of the κ-positive colonies also expressed λ light chains, while $\sim$60% of the λ-positive colonies were also positive for κ light chains. It is important to emphasize that the methodology employed does not distinguish among κ and λ alleles. Thus, many of the "κ only" colonies may also have been derived from light chain-uncommitted cells whose progeny subsequently rearranged both κ alleles. While the same reasoning also applies to "λ only" colonies, the low frequency of successful λ rearrangements observed makes it unlikely that more than a negligible percentage of these clones are in fact expressing two independently rearranged λ genes. More likely is that "λ only" colonies are derived from clonable pre-B cells which were restricted to λ expression at the initiation of culture. Assuming that the likelihood of κ expression is about 10 times as great as the likelihood of λ expression, we have estimated the number of colonies which arise from κ-committed progenitor cells based upon the number of "λ only" colonies and the $\kappa:\lambda$-ratio (SAUTER and PAIGE 1986, to be published). According to these assumptions the number of light-chain-restricted precursors would constitute 43%, 31%, 33%, and 54% of the clonable pre-B cell pool when derived from the fetal liver on the 13th, 14th, 15th, and 16th day of gestation respectively.

There also exist clonable pre-B cells that are uncommitted to a light chain isotype. Based upon the considerations discussed previously, we estimate that they represent approximately 57%, 69%, 67%, and 46% of the clonable pre-B cell pool in fetal liver derived from 13-, 14-, 15-, and 16-day embryos respectively. This interpretation is of course dependent upon the assumption that colonies, even those producing both κ and λ, are derived from single cells. Several lines of evidence suggest that this is the case: (1) Least-squares regression analysis of data obtained in titration experiments indicates clearly that the assay is limiting for a single component, whether measured by μ secretion or by the double production of κ and λ chains (SAUTER and PAIGE 1986, to be published). (2) An accessory cell is not limiting (PAIGE et al. 1985a). (3) The detection of double ($\kappa+\lambda$)-producing colonies might be the result of occasional overlapping of two colonies, each derived from a single precursor cell. We estimated the likelihood of this event based on the total area available for colony formation and the area occupied by a single colony. We found that the number of double-producing colonies detected far exceeded the number expected due to overlap (SAUTER and PAIGE 1986, to be published). (4) The possibility that the limiting unit in the assay was not a single cell, but a doublet consisting of one precursor cell committed to κ expression and one precursor cell committed to λ expression, was excluded by irradiation experiments. Before being cloned in agar, fetal liver cells were exposed to an irradiation dose which caused a reduction of 80%–90% in the number of μ-secreting colonies. If doublets of cells had generated the ($\kappa+\lambda$)-secreting colonies, the observed reduction in the number of ($\kappa+\lambda$)-secreting colonies should have been the square of the reduction of μ-producing colonies. Since we observed that the reduction in the number of μ-producing and ($\kappa+\lambda$)-producing colonies was the same, it follows that the unit generating ($\kappa+\lambda$)-producing colonies is a single light chain-uncommitted precursor cell.

The analysis of light chain expression in fetal liver-derived colonies has revealed two subsets of cells regarding subsequent light chain expression: light chain-unrestricted and -restricted precursor cells. The ratio of light chain-restricted to -unrestricted cells remains relatively constant from day 13 to day 15 of development, whereas on day 16 of gestation a shift to the light chain-restricted phenotype occurs. This suggests that the precursor population in 13- to 15-day fetal liver shows similar differentiative potentials in utilizing light chains genes. In contrast to this, the majority of precursors from day 16 are already committed to a particular light chain gene. This perhaps reflects the progressive transition to a cellular phenotype, which is restricted in its differentiation options. Consistent with this interpretation is the finding that clonable pre-B cells from 16-day fetal liver require only 6–7 days of culture to reach maximum secretion, whereas precursors from 13- to 15-day fetal liver require 8–9 days of incubation (PAIGE et al. 1984). This is further supported by the growth kinetics of mature B cells, which require only 4–5 days of culture to generate Ig-secreting colonies. From this interpretation it follows that a precursor cell which has chosen a particular light-chain gene does not necessarily express this gene immediately. This could imply that the cell needs further signals after commitment or rearrangement for subsequent expression of the rearranged gene. Precedent for such a developmental stage is the pre-B cell line 7OZ/3, which has a functionally rearranged κ light chain gene but expresses it only after induction (Paige et al. 1978; Perry and Kelley 1979; Maki et al. 1980).

The high frequency of light chain-restricted precursor cells is unexpected and seems to be contradictory to the results obtained by the molecular analysis of Abelson-virus-transformed pre-B cell lines from the same developmental age (Alt et al. 1981, 1984). However, there are several factors which make it difficult to compare directly clonable pre-B cells and Abelson-virus-transformed pre-B cell lines. It is not known, for example, whether clonable pre-B cells, especially light chain-restricted cells, are targets for Abelson virus transformation. The clonable pre-B cell assay relies on the functional properties of precursor cells and only detects such cells which can give rise to mature progeny. Therefore, the cell populations studied in these two systems might be different.

Thus, the clonable pre-B cell assay provides culture conditions which allow single precursor cells to differentiate along the B-cell pathway into Ig-secreting cells. Since it is possible to assess the developmental stage of the cell initiating growth by analyzing its progeny, this offers the opportunity to study the emergence of B-cell precursors at different developmental stages and to begin to characterize the conditions required for the transition along this pathway.

References

Abramson S, Miller RG, Phillips RA (1977) The identification in adult bone marrow of pluripotent and restricted stem cells of the myeloid and lymphoid systems. J Exp Med 145:1567–1579

Alt FW, Rosenberg N, Lewis S, Thomas E, Baltimore D (1981) Organization and reorganization of immunoglobulin genes in A-MuLV-transformed cells: rearrangement of heavy but not light chain genes. Cell 27:381–390

Alt FW, Yancopoulos GD, Blackwell TK, Wood C, Thomas E, Boss M, Coffman R, Rosenberg N, Tonegawa S, Baltimore D (1984) Ordered rearrangement of immunoglobulin heavy chain variable region segments. EMBO J 3:1209–1219

Burrows P, Lejeune M, Kearney JF (1979) Evidence that murine pre-B cells synthesize μ heavy chains but no light chains. Nature 280:838–841

Coffman RL, Weissman JL (1981) B220: a B-cell specific member of the T200 glycoprotein family. Nature 289:681–683

Coffman RL, Weissman JL (1983) Immunoglobulin gene rearrangement during pre-B cell differentiation. J Mol Cell Immunol 1:31–38

Coleclough C, Perry RP, Karjalainen K, Weigert M (1981) Aberrant rearrangements contribute significantly to the allelic exclusion of immunoglobulin gene expression. Nature 2390:372–378

Dick JE, Magli MC, Huszar D, Phillips RA, Bernstein A (1985) Introduction of a selectable gene into primitive stem cells capable of long-term reconstitution of the hemopoietic system of W/WV mice. Cell 42:71–79

Giri JG, Kincade PW, Mizel SB (1984) Interleukin-1 mediated induction of κ-light chain synthesis and surface immunoglobulin expression of pre-B cells. J Immunol 132:223–228

Grützmann R (1981) Vergleichende idiotypische Analyse von Rezeptoren mit Spezifität für Histocompatibilitätsantigene. Ph D dissertation, University of Cologne

Haughton G, Lanier LL, Babcock GF (1978) The murine Kappa light chain shift. Nature 275:154–157

Hoffmann MK (1980) Macrophages and T cells control distinct phases of B cell differentiation in the humoral immune response in vitro. J Immunol 125:2076–2081

Keller G, Paige C, Gilboa E, Wagner E (1985) Expression of a foreign gene in myeloid and lymphoid cells derived from multipotent haematopoietic precursors. Nature 318:149–354

Kessler S, Kim KY, Sher I (1981) Surface membrane κ and λ light chain expression on spleen cells of neonatal and maturing normal and immune-defective CBA/N-mice: The $\kappa:\lambda$-ratio is constant. J Immunol 127:1674–1678

Kincade PW (1981) Formation of B lymphocytes in fetal and adult life. Adv Immunol 31:177–245

Kincade PW, Ralph P and Moore MAS (1976) Growth of clones in semi-solid culture is mitogen dependent. J Exp Med 143:1265–1269

Kincade PW, Lee G, Paige CJ, Scheid MP (1981a) Cellular interactions affecting the maturation of murine B lymphocyte precursors in vitro. J Immunol 127:255–260

Kincade PW, Lee G, Watanabe T, Sun L, Scheid M (1981b) Antigens displayed on murine B lymphocyte precursors. J Immunol 127:2262–2268

Maki R, Kearney J, Paige C, Tonegawa S (1980) Immunoglobulin gene rearrangement in immature B cells. Science 209:1366–1369

McGuire KL, Vitetta ES (1981) κ/λ shifts do not occur during maturation of murine B cells. J Immunol 127:1670–1673

McKearn JP, Baum C, Davie J (1984) Cell surface antigens expressed by subsets of pre-B cells and B cells. J Immunol 132:332–339

Melchers F (1977a) B lymphocyte development in fetal liver. I. Development of reactivities to B cell mitogens in vivo and in vitro. Eur J Immunol 7:476–481

Melchers F (1977b) B lymphocyte development in fetal liver. II. Frequencies of precursor B cells during gestation. Eur J Immunol 7:482–487

Melchers F (1979) Murine embryonic B lymphocyte development in the placenta. Nature 277:219–221

Melchers F, Abramczuk J (1980) Murine embryonic blood between days 10 and 13 of gestation as a source of immature precursor B cells. Eur J Immunol 10:763–767

Metcalf D, Nossal GVJ, Warner NL, Miller JFAP, Mandel TE, Layton JE, Gutman GA (1975) Growth of B lymphocyte colonies in vitro. J Exp Med 142:1534–1549

Nishikawa S, Kina T, Gyotoku J, Katsura Y (1984) High frequency of λ gene activation in bone marrow pre-B cells. J Exp Med 159:617–622

Owen JJT, Raff MC, Cooper MD (1975) Studies on the generation of B lymphocytes in the mouse embryo. Eur J Immunol 5:468–473

Paige CJ (1983) Surface immunoglobulin-negative B-cell precursors detected by formation of antibody-secreting colonies in agar. Nature 302:711–713

Paige CJ, Skarvall H (1982) Plaque formation by B cell colonies. J Immunol Methods 52:57–61

Paige CJ, Kincade PW, Ralph P (1978) Murine B cell leukemia line with inducible surface immunoglobulin expression. J Immunol 121:641–647

Paige CJ, Kincade PW, Moore MAS, Lee G (1979) The fate of fetal and adult B-cell progenitors grafted into immunodeficient CBA/N mice. J Exp Med 150:548–563

Paige CJ, Kincade PW, Shinefeld LA, Sato VL (1981) Precursors of murine B lymphocytes. J Exp Med 153:154–165

Paige CJ, Gisler RH, McKearn JP, Iscove NN (1984) Differentiation of murine B cell precursors in agar culture. Frequency, surface marker analysis and requirements for growth of clonable pre-B cells. Eur J Immunol 14:979–987

Paige CJ, Skarvall H, Sauter H (1985a) Differentiation of murine B cell precursors in agar culture. II. Response of precursor-enriched populations to growth stimuli and demonstration that the clonable pre-B cell assay is limiting for the B cell precursor. J Immunol 134:3699–3704

Paige CJ, Skarvall H, Sauter H, Magasiny S (1985b) Development of B cell progenitors in semi-solid agar cultures. In: Lefkovits I, Pernis B (eds) Immunological Methods Vol 3. Academic Press, New York, pp 221–233

Paige CJ, Sauter H, Wu GE (1986) Growth of B cell colonies in double-layer agar cultures. Methods Enzymol, to be published

Palacios R, Henson G, Steinmetz M, McKearn JP (1984) Interleukin 3 supports growth of mouse pre-B cell clones in vitro. Nature 309:127–131
Perry RP, Kelley DE (1979) Immunoglobulin messenger RNAs in murine cell lines that have characteristics of immature B lymphocytes. Cell 18:1333–1339
Perry RP, Kelley DE, Coleclough C, Kearney JF (1981) Organization and expression of immunoglobulin genes in fetal liver hybridomas. Proc Natl Acad Sci USA 78:247–251
Phillips RA, Melchers F (1976) Appearance of functional lymphocytes in fetal liver. J Immunol 117:1099–1103
Sato H, Boyse EA (1976) A new alloantigen expressed selectively on B cells: the Lyb-2 system. Immunogenetics 3:565–572
Shinefeld LA, Sato UL, Rosenberg NW (1980) Monoclonal rat anti-mouse brain antibody detects Abelson murine leukemia virus target cells in mouse bone marrow. Cell 20:11–17
Subbarao B, Mosier DE (1983) Induction of B lymphocyte proliferation by monoclonal anti-Lyb-2 antibody. J Immunol 136:2033–2037
Sugiyama H, Akira S, Yoshida N, Kishimoto S, Yamamura Y, Kincade P, Honjo T, Kishimoto T (1982) Relationship between the rearrangements of immunoglobulin genes, the appearance of a B lymphocyte antigen and immunoglobulin synthesis in murine pre-B cell lines. J Immunol 126:2793–2797
Takemori T, Rajewsky K (1981) Lambda chain expression at different stages of ontogeny on C57BL/6, BALB/c and SJL mice. Eur J Immunol 11:618–625
Teale JM, Mandel TE (1980) Ontogenetic development of B lymphocyte function and tolerance susceptibility in vivo and in vitro fetal organ culture system. J Exp Med 151:429–445
Tonegawa S (1983) Somatic generation of antibody diversity. Nature 302:576–581
Tonegawa S, Maxam AM, Tizard B, Gilbert W (1978) Sequence of a mouse germ-line gene for a variable region of an immunoglobulin light chain. Proc Natl Acad Sci USA 75:1485–1489
Tyan ML, Herzenberg LA (1968) Studies on the ontogeny of the mouse immune system. II. Immunoglobulin-producing cells. J Immunol 101:446–451
Valbuena O, Marcu KB, Weigert M, Perry RP (1978) Multiplicity of germline genes specifying a group of related mouse κ chains with implications for the generation of immunoglobulin diversity. Nature 276:780–784
Wu AM, Till JE, Siminovitch L, McCulloch EA (1968) Cytological evidence for a relationship between normal hematopoietic colony-forming cells and cells of the lymphoid system. J Exp Med 127:455–463
Wysocki LJ, Sato VL (1978) "Planning" for lymphocytes: A method for cell selection. Proc Natl Acad Sci USA 75:2844

Expression of the Immunoglobulin Heavy-Chain Variable Gene Repertoire

B.A. Malynn[1], J.E. Berman[1], G.D. Yancopoulos[1], C.A. Bona[2], and F.W. Alt[1]

1 Introduction

A major challenge in immunology is the elucidation of the cellular and molecular mechanisms which determine the expressed antibody repertoire. A large body of knowledge concerning murine antibody gene organization and diversity has been accumulated. However, the underlying processes governing usage of these genes remains in question. Knowledge of the mechanisms which regulate immunoglobulin variable (V) gene rearrangement and the relative expression of the various V gene families in developing B-cell populations may be crucial to an understanding of generation of the antibody repertoire in normal as well as

[1] Department of Biochemistry and Molecular Biophysics, Columbia University College of Physicians and Surgeons, 630 West 168 Street, New York, NY 10032, USA
[2] Department of Microbiology, Mount Sinai School of Medicine, New York, NY, USA

Current Topics in Microbiology and Immunology, Vol. 135

disease states. Current investigations of human V_H gene organization and diversity will eventually allow extension of information gained about the murine repertoire to studies of V_H gene usage in humans.

2 Structure and Organization of Mouse V_H Genes

2.1 Immunoglobulin Molecules

Immunoglobulin (Ig) molecules are composed of a common polymeric structure consisting of two identical heavy and two identical light polypeptide chains. Each chain is composed of an N-terminal variable region and a C-terminal constant region. The primary determinants of antigen-binding specificity reside in the variable region, while the constant region is involved with effector functions (complement fixation, placental transfer, etc.). The variable region of heavy and light chains can be further subdivided into regions of relatively conserved amino acid sequence (framework regions) and regions of highly variable amino acid sequence, referred to as hypervariable regions or complementarity determining regions (CDRs). The CDRs are thought to be areas of the protein directly involved with antigen contact. The variable regions of both the heavy and light chain genes are assembled from component gene segments. The heavy-chain variable region gene is assembled from three different elements: the variable segment (V_H) which encodes for the (approximately) 98 N-terminal amino acids, the diversity segment (D) which encodes the next 1–15 amino acids, and the joining segment (J_H) which encodes the final (approximately) 12–17 amino acids; families of these segments are encoded in separate clusters along the chromosome (for review see TONEGAWA 1983). Formation of a functional heavy-chain variable region gene is one of the first recognized events in B-cell development and involves at least two recombinational events: both joining of a D segment to a J_H segment and joining of a V_H segment to the pre-existing DJ_H complex are needed to form the complete V_HDJ_H heavy-chain variable region gene. The gene segments encoding the light-chain variable region are assembled subsequent to the assembly and expression of a functional heavy chain. This assembly event for light chains of either the κ or λ type involves the joining of a variable gene segment (V_L) to a joining segment (J_L) to form a complete V_LJ_L variable region; no D segments are involved in light-chain variable regions. For both heavy- and light-chain genes, transcription of the mRNA is initiated upstream of the V region and terminated downstream of the C region; intervening sequences are removed by standard RNA processing mechanisms (for review of Ig gene assembly and expression, see YANCOPOULOS and ALT 1986).

2.2 Generation of Diversity

The diversity of antigen-binding specificities is generated at multiple levels. The germ-line variable region elements (V, D, and J) occur in multiple, distinct copies which encode different primary amino acid sequences, particularly in the CDRs. Current evidence suggests a lack of any general restriction on the combinatorial assortment of various *V*, (*D*), and *J* elements (TONEGAWA 1983) (although this issue remains open), thus providing a source of diversity which results from "combinatorial" association of the different germ-line elements. The portions of the heavy- and light-chain variable regions encoded by the $V_H D J_H$ or $V_L J_L$ junction are involved with antigen contact (CDR3); significantly, junctional diversification mechanisms operative at the time of *V* gene assembly further diversify the coding potential of this important region. Thus, bases may be lost from the potential coding portions of the V_H, *D* and J_H segments and from the V_L and J_L segments; furthermore, nucleotides can be added *de novo* to the V_H to *D* and *D* to J_H interfaces (TONEGAWA 1983; ALT et al. 1986). This "junctional" diversity can alter the amino acid sequence of CDR3 and thus be an important somatic source of antibody diversity. Thus, although any given B-cell clone assembles only one functional heavy- and light-chain gene, the total number of different *V* genes which can be assembled somatically by the B-cell population in an animal appears to be immense due to the germ-line, combinatorial, and junctional diversification mechanisms. Furthermore, the antigen-binding pocket of an antibody molecule is generated from the interaction of the variable regions of the heavy and light Ig chains; thus, the combinatorial association of heavy and light chains having different V regions further increases the diversity of antibody specificities. Finally, comparison of the nucleotide sequences of V_H or V_L genes expressed by mature Ig-secreting cell lines to those of their germ-line counterparts has provided strong support for the involvement of somatic hypermutational mechanisms in the generation of diversity. Hypermutation is most evident in the CDRs, which is consistent with an antigen-mediated selection process (e.g., BOTHWELL et al. 1981; CREWS et al. 1981). In general, it appears that somatic mutation is much more pronounced later in the immune response (GEARHEART et al. 1981; CREWS et al. 1981), although the nature of the mechanism or the process by which it is induced is not clear.

2.3 Organization of *V* Gene Families

The best-characterized variable gene locus is that of the murine heavy-chain genes. In general, little is known about the organization and complexity of the families of gene segments which encode the variable regions of human Ig heavy and light chains. Furthermore, in both human and murine species, much more is known about the organization of the heavy-chain loci than the light-chain loci. In the mouse genome, a cluster of four J_H segments lies 7 kb upstream from the μ constant region exon (C_μ). Approximately 10 *D* segments are located 1–80 kb upstream from the J_H cluster (WOOD and TONEGAWA 1980). An estimated 100–1000 V_H segments lie at an unknown distance upstream from the *D* cluster. These V_H segments have been subdivided into nine distinct families, ranging in size from a few to more than 60 members (BRODEUR and RIBLET 1984; BRODEUR et al. 1984; DILDROP 1984; WINTER et al. 1985) (Table 1, second

Table 1. Comparison of human and murine V_H gene families

Murine V_H subgroups[a]	Murine V_H families[b]	Murine V_H complexity[c]	Homologous human family	Human V_H complexity[d]	V_H subsets
I	Q52	15	II?	7–8	–
	36–60	5			
II, V	J558	60	Subgroup I[e]	5–7	J558 homol.
–			–	8–10	J558 nonhomol.
III	S107	4	Subgroup III[f]	3–4	S107/J606
	J606	10		3–4	homol.
	441-4	2		18–20	441-4/7183
	7183	12		18–20	homol.[g]

[a] As defined by KABAT et al. (1983) using amino acid sequence homology
[b] As defined by BRODEUR and RIBLET (1984) using nucleic acid sequence homology
[c] The estimated number of genes per family (a minimum estimate based on the number of fragments detected by Southern analysis, BRODEUR and RIBLET 1984)
[d] The estimated number of cross-reacting human *Hin*dIII fragments seen by Southern blotting analysis. In subgroup III, the murine V_H probes identify overlapping human *Hin*dIII fragments
[e] Subgroup I was previously identified using a V_H104 probe, a member of the J558 family (RECHAVI et al. 1983)
[f] Subgroup III was previously identified using an S107 probe (RECHAVI et al. 1982)
[g] This subset includes the S107 and J606 homologous genes

and third columns). In general, genes within a V_H family share greater than 80% nucleic acid sequence homology, while genes from different families are less than 70% homologous (BRODEUR and RIBLET 1984; BRODEUR et al. 1984). In the few murine strains examined, V_H families are often organized in nonoverlapping clusters with individual members of a family separated by an average distance of 10–20 kb (KEMP et al. 1981; BOTHWELL et al. 1981; LOH et al. 1983). The distance between families remains unknown; however, using deletion and recombinant-inbred strain mapping the order of V_H families in several strains has been determined as, from 5′ to 3′: centromere, V_HJ558, V_HS107, V_HQ52, V_H7183, *D* cluster, J_H segments, constant-region genes. V_H36–60, V_H441–4, and V_HJ606 are located in an undetermined order upstream from V_HJ558 (BRODEUR et al. 1984). In strains examined to date, the V_HJ558 family is by far the largest, containing as many or more members than the other families combined. Recent data suggest that the order of the families defined in BALB/c mice may not be the same in all mouse strains. In NIH/Swiss mice, the V_HQ52 family, or at least some of its members, are located 3′ of the V_H7183 genes (RETH et al. 1986a).

In mice there are four functional κ light chain *J* segments and at least 200 V_κ segments grouped into families similar to those of the V_H segments (CORY et al. 1981; SELSING and STORB 1981). The murine λ locus is organized in a somewhat different fashion, with only two V_λ genes, each of which is followed downstream by a J_λ unit (EISEN and REILLY 1985). The λ locus is more complex in humans where λ comprises a larger fraction of the serum light chain (for review see YANCOPOULOS and ALT 1986). It is obvious that in order to be able fully to understand the development and nature of the antibody repertoire, a detailed understanding and characterization of both the heavy- and the light-chain loci will be necessary. However, due to the wealth of information concerning heavy-chain variable region gene structure and function and the relative paucity of information concerning that of light chains, the remainder of this review will focus on the development of the V_H repertoire.

3 Regulation of the Assembly and Expression of Ig Heavy-Chain Variable Regions

3.1 V_H Gene Assembly in Pre-B Cell Lines

Fetal and newborn liver and adult bone marrow are the sites of B-cell differentiation in vivo. Transformation of cells from these hematopoietic organs with Abelson murine leukemia virus (A-MuLV) generates cell lines representative of the early stages of B-cell development (BALTIMORE et al. 1979). Fetal liver-derived lines generally represent the earliest defined B-cell precursor; that is, cells which do not produce a complete Ig heavy chain, but contain DJ_H recombinations on both chromosomes. As these cells are propagated in culture they actively rearrange V gene segments to yield complete V_HDJ_H complexes and some subsequently express cytoplasmic μ chain. Many lines derived from adult bone marrow had made complete V_HDJ_H rearrangements at the time of isolation (ALT et al. 1984). Few A-MuLV transformants from any source have rearranged light-chain genes at the time of isolation, although several rearrange light-chain genes during propagation in culture (LEWIS et al. 1982; RETH et al. 1985). At least one transformant which had DJ_H rearrangements on both chromosomes upon isolation undergoes all of the sequential Ig gene assembly and expression events in culture which ultimately lead to the generation of surface IgM$^+$ progeny (RETH et al. 1985).

3.2 Expression of Germ-Line or Incompletely Assembled V Gene Segments

The heavy-chain loci in the earliest (dual DJ_H rearranged) B-lineage cells are not transcriptionally silent; transcription of variable and constant-region gene segments takes place even before complete V_HDJ_H rearrangement has occurred (ALT et al. 1982; RETH and ALT 1984; LENNON and PERRY 1985). Promoters upstream of D segments employed in DJ_H rearrangements are transcriptionally active; perhaps mediated by the increased proximity to the heavy-chain enhancer. The so-called D_μ mRNAs which result from such transcriptional activity can encode D_μ proteins which contain a complete C_μ region but lack a classical variable region; these proteins do have a variable N terminus encoded by D and J_H sequences. Certain characteristics of the D_μ proteins suggest some, as yet undetermined, function (RETH and ALT 1984).

Expression of germ-line (unrearranged) V gene segments occurs independently of the classical heavy-chain enhancer and is restricted to the very early stages of B-lymphocyte differentiation, predominantly in cells or tissues in which heavy-chain genes are actively rearranging. To date, very high level expression has only been observed for members of the V_HJ558 family; several types of analyses suggest that many members of this very large V_H gene family are expressed (YANCOPOULOS and ALT 1985). Although the structure of germ-line variable region "mRNA" sequences suggests that they could encode proteins, this remains to be demonstrated in B cells. Surface expression of a germ-line V_H protein was reported to occur in a T-cell line (UCKER et al. 1985). Any function for such proteins remains highly speculative; however, it is worth noting that "V proteins" would resemble other single Ig-domain molecules such as β_2-microglobulin and therefore would have the potential of being expressed on the cell surface in association with other molecules (e.g., MHC class I molecules). Germ-line C_κ and V_λ transcripts have also been detected in cell lines representing later stages of B-cell differentiation (VAN NESS et al. 1981; PICARD and SCHAFFNER 1984; NELSON et al. 1985). Perhaps the most significant finding with respect to the expression of unrearranged V gene segments is a correlation between transcriptional activity and rearrangement. A variety of recent experiments suggest a correlation between the "openness" of the V_H locus and its accessibility to recombinational, as well as transcriptional, machinery; regulation of rearrangement events in lymphocyte differentiation appears to be controlled at the level of accessibility of appropriate Ig gene segments to putative recombinase enzyme(s) (reviewed in ALT et al. 1986).

4 Expression of the Murine V_H Repertoire

4.1 Background

Studies in vivo and in vitro have suggested that reactivity to certain antigens appears in an ordered fashion during ontogeny (SILVERSTEIN et al. 1963; PRESS and KLINMAN 1975; SHERWIN and ROWLANDS 1975; LYDYARD et al. 1976; SIGAL et al. 1976; SIGAL 1977; CANCRO et al. 1979; DENIS and KLINMAN 1983; STOHRER and KEARNEY 1984; TEALE 1985). For example, in BALB/c mice there is a sequential appearance of B cells responsive to dinitrophenyl, trinitrophenyl, fluorescein, nitrophenyl, and phosphorylcholine (KLINMAN and PRESS 1975; TEALE 1985). These phenomena could result from an ordered program of V_H (or V_L) gene utilization in ontogeny and/or from various types of cellular selection processes. Alternatively, the utilization of the various V genes segments could be random and/or not play a significant role in the ordered appearance of particular specificities.

4.2 V_H Gene Utilization by Pre-B Cell Lines

The availability of A-MuLV-transformed, surface-Ig-negative pre-B cell lines which actively perform V_H to DJ_H joining during propagation in culture provided a unique opportunity to study V_H utilization patterns in the absence of the antigenic or immunoregulatory forces which act on pre-B cells differentiating in vivo. Analyses of such lines derived from the fetal liver of BALB/c mice revealed that V_H utilization is not random, but is characterized by the overwhelming usage of members from the V_H7183 gene family and, in particular, one member of this family, V_H81X (YANCOPOULOS et al. 1984). Similar analyses have indicated that fetal liver hybridomas and A-MuLV-transformed pre-B cells (predominantly from BALB/c mice, but from other strains as well) which had assembled V_HDJ_H rearrangements at the time of isolation (surface-Ig-negative usually due to lack of light-chain rearrangement) also preferentially utilized V_H7183 segments in V_HDJ_H rearrangements (YANCOPOULOS et al. 1984; PERLMUTTER et al. 1985). As indicated above, the V_H7183 gene family was demonstrated to be located most 3′ (J_H-proximal) in BALB/c mice; furthermore, V_H81X was mapped as the most J_H-proximal (functional) member of this family (YANCOPOULOS et al. 1984). To date, the V_H81X gene has not been reported to encode the expressed heavy chain of any Ig-secreting cell. In addition, analysis of the V_H81X sequence suggested certain potential features of a V_H81X polypeptide which would impair interaction with a light-chain V region (RETH, unpublished observations), although this has not been directly demonstrated. Furthermore, the V_H81X gene has been found predominantly, if not exclusively, in nonproductive rearrangements in BALB/c fetal-liver-derived, A-MuLV-transformed pre-B lines, suggesting a possible selective disadvantage to transformants which rearranged this (and perhaps other) V_H genes productively. Thus, the functional significance of high-frequency V_H81X rearrangement seems questionable. This interpretation is further emphasized by studies of an A-MuLV transformant derived from the adult marrow of a NIH/Swiss outbred mouse. Deletion mapping analyses of the most 3′ V_H families in this line indicated interspersion of the V_HQ52 and V_H7183 families; in particular, this line did not appear to have a V_H81X equivalent and at least a portion of the V_HQ52 family occurred 3′ to V_H7183 family members. Strikingly, this line preferentially employed V_HQ52 members in its V_H to DJ_H rearrangements (RETH et al. 1986a). The most simple interpretation of the rearrangement patterns in these lines is that due to mechanistic aspects of the recombinase system, the frequency with which particular V_H segments are utilized in V_H to DJ_H rearrangements is directly related to J_H proximity, with more proximal (3′) segments being rearranged more frequently. It should, however, be noted that other interpretations of these data exist. For example, the NIH/Swiss line was derived from adult marrow, whereas all BALB/c lines which actively rearrange heavy-chain genes that have been studied thus far were derived from fetal liver. Thus, the frequency with which different families of V_H segments are utilized may change during the maturation of a particular pre-B lineage (PERLMUTTER et al. 1985; RETH et al. 1986a; see below for further discussion of this point). Clearly, final resolution of such questions may be provided by more extensive analyses of actively rearranging A-MuLV transformants or, perhaps more definitively, by analyses of V_H utilization patterns of in vivo or in vitro differentiated pre-B cell clones (WHITLOCK and WITTE 1982; PAIGE 1983; PALACIOS et al. 1984; SPALDING and GRIFFIN 1986, from various strains, at various stages of development, and under the influence of various types of accessory cells.

It was also notable that a high proportion of the V_HDJ_H rearrangements formed during culture of the NIH/Swiss-derived lines were productive. Thus, the differentiation of these pre-B cell lines to later stages of development proceeded much more efficiently than was observed with the pre-B lines derived from BALB/c fetal liver (which rarely generated progeny with a productive V_HDJ_H rearrangement). Although the reason for the relative abilities of the two different types of lines to generate progeny which produce μ heavy chains is unknown, one possibility could be some relationship to the predominant V_H genes which are rearranged in the lines. If this interpretation is correct, then it follows that the relative efficiency of pre-B cell differentiation in the two strains may also reflect V_H gene organization (RETH et al. 1985; 1986; ALT et al. 1986). Such possibilities remind us that potential ambiguities may not only arise from studies of cultured or transformed cells, but also be associated with the use of inbred mouse lines.

4.3 V_H Gene Utilization In Vivo

Despite the frequent usage of V_H7183 members in pre-B cell lines and fetal liver hybridomas, members of this family, in general, are rarely found in the V_HDJ_H joins which have been characterized in mature B-cell tumors (KABAT et al. 1983) and to date have not been frequently reported to be used in murine responses to conventional antigens. In particular, usage of V_H81X has thus far not been reported in functional rearrangements. These findings raised the possibility that the biased usage of particular V segments in pre-B cell lines might not reflect mechanisms operating during the differentiation of normal pre-B cells in vivo. To address this issue, we compared the expressed V_H repertoire of central and peripheral B-cell differentiation organs at various stages of development and in different mouse strains (BLACKWELL et al. 1984; YANCOPOULOS et al. 1986c; MALYNN et al., in preparation) (Table 2). For these studies we prepared a panel of V_H probes specific for six of the seven well-characterized V_H gene families which were used in a Northern blotting assay to determine the expressed V_H repertoire (Table 2) (YANCOPOULOS et al. 1986c). The basis for this assay was the previous finding that the absolute level of μ mRNA expression is similar among cells of the same differentiation stage and is not related to the particular V_H segment expressed. Thus, in a population of B-lineage cells which are at the same stage of differentiation but contain heterogeneous V_HDJ_H rearrangements, the relative proportion of μ mRNA containing a particular V_H should constitute a rough estimate of the frequency within the population of V_HDJ_H rearrangements employing that V_H.

To estimate the absolute fraction of V_HDJ_H rearrangements utilizing V_H81X and other members of the V_H7183 family in a given population of cells, we compared the relative hybridization of a V_H81X-specific probe to a standardized amount of μ mRNA from various sources. As a control, we employed RNA from cell lines which produced only V_H7183- or V_H81X-containing μ mRNA; thus, all of the μ mRNA from these lines should hybridize to a V_H7183 probe (BLACKWELL et al. 1984; YANCOPOULOS et al. 1986c). Strikingly, a comparison

Table 2. Utilization of V_H gene segments during B-cell development in normal strains of mice

V_H Family	BALB/c[a]	C57BL/6[a]	C3H/HeJ[b]	CBA/J[b]
7183 (V_H81X)	12.4(4.0)[c]	6.1	3.1	4.3
7183 (V_HD6.96)	4.4(1.8)	2.7	5.4	2.5
Q52	1.3(0.9)	1.5	nd	0.5
S107	0.9(1.0)	0.6	1.6	0.3
J558	0.4(0.2)	0.2	0.5	0.2
J606	<0.2(0.2)	<0.2	nd	nd
36–60	1.2(0.8)	nd	nd	nd

Replicate Northern blots containing total RNA from adult spleens or poly-A$^+$ RNA from newborn livers from the indicated mouse strains were probed with either a C_μ probe or with probes specific for the V_H families as shown. The signal intensities were quantitated by densitometry, and preference values (P) were calculated according to the following equation:

$$P = \frac{\text{relative use in newborn liver}}{\text{relative use in adult spleen}}$$

nd, not determined

[a] From YANCOPOULOS et al. (1986c)
[b] From MALYNN et al. (in preparation)
[c] The preference values comparing the relative usage in newborn livers to that of LPS-stimulated spleen are indicated in parentheses

of the standardized hybridization ratio of V_H81X to μ mRNA of fetal or newborn liver (a primary B-cell differentiation organ rich in pre-B cells and immature B lymphocytes) with that of lines which made V_H7183-containing μ mRNA demonstrated that V_H81X and closely related genes are utilized on a substantial proportion of the μ mRNA sequences of fetal liver B cells (15%–20% or more). A similar analysis indicated that the V_H7183-containing RNA sequences comprised a much smaller proportion of the μ mRNA of resting (less than 2%) or LPS-stimulated (less than 5%) adult spleen (which is a peripheral lymphoid organ that contains substantial populations of mature B-lymphocytes and some Ig-secreting plasma cells). LPS polyclonally activates a large fraction of B cells in spleen; thus, the μ mRNA from such cells should now be more representative of the expressed repertoire of the entire splenic B-cell population rather than that of plasma cells, whose high levels of μ mRNA expression might influence the results with unstimulated spleen populations. Together, these results indicated that the V_H7183 segments are utilized in a substantial proportion of the rearrangements in the newly differentiating B-lymphoid cells of the fetal liver, but that the contribution of this family to the expressed rearrangements in the more mature B-lymphoid cells of the spleen is greatly reduced (YANCOPOULOS et al. 1986c). In particular, these findings support the indications, from the studies of pre-B cell lines, that V_H7183 segments are preferentially rearranged in differentiating pre-B cells of certain mouse strains.

To further elucidate the apparent shift between the fetal and the adult repertoire, we compared the relative representation of each of the V_H families in the μ mRNA of fetal liver cells with that of adult spleen (YANCOPOULOS et al. 1986c; MALYNN et al., in preparation) (Table 2). Replicative Northern blots containing RNA prepared from newborn livers or adult spleen were probed either with C_μ or with probes representative of each of the V_H families. The signal intensities were quantitated and the relative expression of each of the V_H gene families in newborn vs adult mice was compared by a simple formula which assigns a preference value for the expression of each V_H family (BLACKWELL et al. 1984; YANCOPOULOS et al. 1986c) (Table 2). A preference value of one indicates that a given V_H family is utilized in an equal proportion of the μ mRNA of fetal liver and adult spleen; a value greater than one indicates increased representation in newborn liver relative to adult spleen; a value less than one indicates decreased representation in newborn liver relative to adult spleen. Strikingly, these studies indicated that the preference value of each family (with the possible exception of V_H3660; see below) was directly related to J_H proximity, with the more 3′ families being preferentially represented in newborn liver (Table 2). This position-dependent bias was evident whether the newborn repertoire was compared to resting or LPS-stimulated splenic lymphocytes, although the differences were not as dramatic with the LPS-stimulated cells (see below for further discussion of this point). Based on our studies of the absolute levels of V_H7183 expression in newborn liver relative to adult spleen and the relationship between rearrangement and expression of V_H gene segments in pre-B lines, it seems likely that V_H gene utilization in fetal liver is based on the absolute probability of rearranging a particular V_H gene segment within a V_H gene family. Thus, the absolute utilization of the entire family early in development would be a function of rearrangement probability and family size. For example, the probability of utilization of an individual member of the V_HJ558 family might be low due to position effects, yet the absolute probability of rearranging a member of this family could be relatively high due to the large number of members. A direct relationship between preference values and position would result if one assumes that, although the fetal repertoire is determined mostly by position-dependent rearrangement effects, the repertoire of the resting spleen (a peripheral organ in which rearrangement is presumably not occurring) is "normalized" to be representative of absolute family size. Significantly, analyses of the V_H utilization patterns of a panel of hybridomas derived from LPS-stimulated splenic lymphocytes has provided preliminary evidence of V_H gene utilization that is consistent with frequency corresponding to overall family size (DILDROP et al. 1985). Thus, in the resting spleen, V_HJ558 would account for a large percentage of the μ mRNA (because of the many family members) while smaller families like S107 (one or two functional members) would account for a small percentage. According to this notion, the preference numbers that we have derived would result from dividing the frequency with which a given family is rearranged (a function of family size and map position; the basis for representation in fetal liver) by its total number of members (the proposed basis for representation in adult spleen; see below for further discussion of how normalization might occur), which would give a direct relationship if rearrangement was primarily position-dependent. Given these assumptions and the preliminary data suggesting that representation in splenic lymphocytes is, in fact, correlated with family size (DILDROP et al. 1985; YANCOPOULOS et al. 1986c), the direct relationship between map position and preference values is consistent with the proposal that during normal pre-B cell differentiation in the fetal liver, recombinase works by a one-dimensional tracking mechanism during V_H to DJ_H joining, preferentially recognizing V_H segments more proximal to the J_H locus.

4.4 Potential Mechanisms for Repertoire Normalization

The major shift away from the position-dependent V_H repertoire of the fetus which is observed in B cells of the adult spleen could be effected at various levels, ranging from changes in the program of V_H gene rearrangement in more mature cells to cellular selective forces which might act (either positively or negatively) to normalize the representation of the various V_H gene families in mature B-cell populations. Below, we will consider various of the possible mechanisms; although none of these mechanisms can be excluded on the basis of current evidence, for reasons that will be outlined we feel that normalization of V_H gene expression in adult spleen is due in large part to antigen selection and/or immunoregulatory forces.

4.4.1 Normalization Mechanisms Based upon Cellular Selection

We have demonstrated that the preference values of the V_H7183 and V_HJ558 families were constant from day 14 of fetal life through birth; this finding suggested that throughout fetal development all V_H segments are rearranged and expressed at a constant rate which is dependent on chromosomal position, as opposed to sequential rearrangement of particular families (Yancopoulos et al. 1986c). Furthermore, the position-dependent bias of V_H gene utilization is also evident in A-MuLV transformants derived from adult bone marrow (Yancopoulos et al. 1984) and, in preliminary studies, in populations of bone-marrow-derived B cells (Yancopoulos et al. 1986c), suggesting that repertoire normalization may actually occur in peripheral lymphoid organs. If the results with adult-marrow-derived B cells are confirmed, it seems unlikely that the shift in the V_H repertoire during development is due primarily to a stage (or maturation)-dependent shift in the rearrangement of particular V_H gene families, although such mechanisms (see below) could still play some role (potentially in particular B-cell subsets). Therefore, it seems quite likely that selection plays a major role in repertoire normalization; additional evidence to support this conclusion comes from studies of mice with a severe combined immunodeficiency disease (SCID) (Bosma et al. 1983; Yancopoulos et al. 1986c; see below).

Selection could occur by endogenous or environmental antigens or regulatory cells acting on surface antibody receptors, or by other mechanisms acting even prior to the expression of surface antibody receptors. A potential example of the latter mechanism involves the putative inability of the V_H81X-encoded heavy chains to interact efficiently with most light chains (see above); such a property could drastically reduce the representation of the V_H7183 family in the rearrangements of B cells. In accordance with this possibility, the preference value for V_H81X is higher than that measured with a probe specific for another more representative member of the V_H7183 family. Thus, the observed higher relative expression of V_H81X in newborn liver vs adult spleen may be accounted for not only because of its higher absolute level of expression in newborn liver (shown above) but also because of its possible underrepresentation in adult spleen. With respect to the overall representation of various families, we have observed a very similar V_H repertoire in both unstimulated and LPS-stimulated splenic lymphocytes. The only notable difference in the two spleen populations appears to result from the two- to threefold increase in the absolute expression levels of V_H7183-containing μ mRNA sequences in the LPS-stimulated B cells relative to those of the unprimed spleen (see above). In the context of normalization mechanisms based upon cellular selection, there are several possible explanations, not necessarily mutually exclusive, which might explain this difference. First, resting B cells bearing V_H7183-encoded heavy chains may only rarely be stimulated to the Ig-secreting stage in an unprimed adult spleen, which would predict that this family would be relatively underrepresented in antigen responses. Alternatively, assuming that a large fraction of the LPS-inducible B cells in spleen represent immediate descendants of bone marrow precursors, the increased V_H7183 expression in the LPS-stimulated spleen may reflect preferential V_H7183 expression in pre-B cells in the adult marrow.

4.4.2 Rearrangement Mechanisms to Explain Repertoire Normalization

Potential rearrangement mechanisms that could conceivably account for repertoire normalization at the pre-B stage include programmed changes in the V_H utilization patterns (Perlmutter et al. 1985; Reth et al. 1986) or, at either the pre-B or even the B-cell stage, V gene replacement mechanisms (Yancopoulos et al. 1984); such mechanisms could theoretically be subject to developmental or immunoregulatory modes of regulation. Thus, until the issue of the repertoire of newly arising marrow and splenic lymphocytes is unequivocally resolved, it remains possible that mechanisms later in development work to promote the V_H rearrangements of more 5′ genes in adult bone marrow, for example, perhaps a change in recombinase action from an "associative tracking" to a "dissociative tracking" mechanism (Wood and Tonegawa 1983; Reth et al. 1986a) possibly promoted by the appearance of stage-specific recombinase "entry" sites in more 5′ regions of the V_H locus. V_H gene replacement mechanisms provide an additional potential mechanisms of repertoire normalization. Potentially site-specific replacement of heavy chain V_HDJ_H rearrangements with uncharacterized sequences (Yancopoulos et al. 1986b) or with another V_H sequence (Reth et al. 1986b; Kleinfield et al. 1986) has been observed in cell lines. Replacement of V_LJ_L joins in cell lines via joining of an upstream V_L to a downstream J_L has also been documented (Lewis et al. 1982; Reth and Alt, unpublished data). However, the relevance of these findings with respect to normal physiology remains to be determined (Alt 1986). As one approach to the issue of mechanistic vs cellular selection normalization mechanisms, we are exploring an independent method of estimating the relative contribution of the different *V* gene families to the repertoire of resting vs activated B cells of the adult marrow and resting spleen. This approach involves quantitating the relative contribution of the different V_H gene families to the membrane (primarily in pre-B or B cells) vs the secreted (primarily in plasma cells) form of μ mRNA. Similar types of analyses of the relative expression in μ vs γ mRNA may help detect differences arising during secondary responses.

4.4.3 Preferential V Gene Utilization and Stage-Specific Antibody Responses

Our finding that the utilization of the V_H gene families by differentiating pre-B cells appears to remain constant during fetal (and probably adult) B-cell differentiation (see above) implies that reports of a restricted antibody repertoire early in development (SILVERSTEIN et al. 1963; PRESS and KLINMAN 1975; SHERWIN and ROWLANDS 1975; LYDYARD et al. 1976; SIGAL et al. 1976; SIGAL 1977; CANCRO et al. 1979; DENIS and KLINMAN 1983; STOHRER and KEARNEY 1984; TEALE 1985) cannot be explained exclusively on the basis of the absence of rearrangements utilizing a particular V_H gene segment. However, it seems possible that the great relative bias in the representation of certain V_H gene segments in the newly generated vs normalized repertoires might account for preferential utilization of certain V_H gene segments in fetal as opposed to adult responses: a V_H7183 member closely related to a *V* gene segment found rearranged in pre-B cells is utilized in a characterized early response (RILEY et al. 1985). Although our results on the development of the V_H repertoire suggest that it is unlikely that programmed rearrangement of V_H segments evolved to ensure an ordered program of antibody specificities during ontogeny, ordered V_L rearrangement or other selective mechanisms which lead to such a program are still conceivable. For example, it has recently been demonstrated that strain-specific clonotype silencing, and not strain-specific differences in the V_H segments which are rearranged, are responsible for determining the predominant antibodies used in the antidextran response by different strains (FROSCHER and KLINMAN 1985). It is notable that in our studies, only the V_H36–60 family was found to be utilized at a higher relative level in the neonate than would have been expected from its reported J_H-distal position (YANCOPOULOS et al. 1986c). Notably, this family is used to encode heavy chains utilized in early responses in the animal (RILEY et al. 1985), and is rearranged at a relatively high frequency in pre-B cell lines (YANCOPOULOS et al., unpublished data). Together, these data raise the possibility that either some members of this family are located further J_H-proximal than previously thought, or that mechanisms (rearrangement or cellular selection) exist to increase the frequency of cells expressing V_H36–60 rearrangements early in development.

In conclusion, usage of a particular *V* gene in an immune response is probably the result of a complex balance struck between frequency of V_H gene rearrangement and selection by antigen and T cells. The relative importance of these two factors in normal animals probably varies with different antigens and at different times in the immune response. For example, early in the response a more frequently rearranged *V* gene of lower affinity may predominate at the expense of a less commonly encountered *V* gene of higher affinity. This seems even more plausible in light of the finding that different germ-line genes are utilized in the primary than in the secondary response to oxazalone (Berek et al. 1985). Defects affecting the regulation of the frequency of V_H gene usage or in antigen and T-cell selection might upset the normal equilibrium between the two and result in abnormal responses such as those seen in autoimmune strains.

4.5 V_H Gene Usage in Disease States

4.5.1 Autoimmune Strains

The rearrangement and subsequent expression of various V gene families may provide insights into understanding the generation of the antibody repertoire in disease states such as autoimmunity and immunodeficiency diseases. Analysis of a panel of murine monoclonal autoantibodies with a wide range of antigenic specificities (including thyroglobulin, DNA, Sm, acetylcholine receptor, collagen, RBC, and IgG) revealed a biased utilization of V_H families (Manheimer-Lory et al. 1986; Monestier et al. 1986). Specifically, of 53 autoantibodies tested, 21 used a V_H gene from J558, one from S107, seven from Q52, and 24 from 7183. In this case, relative to the number of V_H genes per family, there is a clear bias for utilization of the most 3′ families, particularly V_H7183. Although the number of genes encoding autoantibodies that have been sequenced to date is limited, the available data suggest that autoantibodies derive from the same germ-line *V* gene pools which encode antibodies to foreign antigens (Manheimer-Lory et al. 1986; Kofler et al. 1985). Sequence analysis of two representative rheumatoid factor (RF) V_H genes failed to identify a common heavy chain variable region gene segment (V_H, *D*, or J_H) which might account for the antigenic specificity and cross-reactive idiotypes shared by these RF (Manheimer-Lory et al. 1986).

We are currently applying the type of analysis described in the in vivo V_H gene utilization studies to examine the developmental pattern of V_H gene expression in several autoimmune strains. Preliminary data indicate that the general pattern of V_H gene expression previously demonstrated for normal strains, that is, highly skewed expression of 3′ family genes early in development, is also observed in neonatal mice of the autoimmune strain, NZB. Examination of the expressed V_H repertoire in various B-cell subsets should provide additional insight into the development of the V_H repertoire in normal, autoimmune, and other immunodefective mouse strains. For example, $Ly1^+$ B cells have been implicated in the production of autoantibodies (Hayakawa et al. 1984) and are present in increased numbers in NZB mice (Hayakawa et al. 1983). Such analyses could also point to the stage at which particular defects may arise. For example, the use of J_H-proximal V_H segments at a higher than predicted frequency in certain autoantibodies (Manheimer-Lory et al. 1986) may reflect not the particular binding specificities of these V_H segments, but rather the B-cell developmental stage or B-cell subset in which a breakdown in normal regulatory processes leads to the development of anti-self immunoglobulins.

4.5.2 Immunodeficient Mouse Strains

In preliminary studies, we have demonstrated that mice carrying a SCID mutation (Bosma et al. 1983) produce a truncated μ mRNA in their adult spleen which appears to contain V_H sequences (Yancopoulos et al. 1986c). Significantly, this apparently aberrant μ mRNA of the adult spleen maintains a V_H repertoire that is similar to the fetal repertoire seen in normal mice. The generation of these aberrant μ mRNA sequences in SCID mice may provide a unique insight into the mechanisms which serve to normalize the V_H repertoire found in the mature B cells of peripheral lymphoid organs. In this regard, the short V_H-hybridizing transcripts found in SCID mice may encode defective heavy-chain proteins that prevent progression of pre-B cells to functional B cells. If repertoire normalization occurs via selection at the B-cell stage, such a defect could explain the lack of B cells and the maintenance of the fetal repertoire in adult SCID mice. Assuming that other conceivable mechanisms for repertoire normalization, such as developmentally programmed V_H gene utilization or V_H gene replacement, would not depend on the production of complete μ chains, then the SCID results would provide strong preliminary evidence that selection plays a major role in repertoire normalization.

5 Organization of V_H Genes in Humans

5.1 Human V_H Gene Families

As outlined above, interpretation of V_H gene usage studies in the mouse have relied on the large available pool of V_H gene sequences and on information concerning their chromosomal organization into homologous families. The human V_H genes have been divided into three families based on amino acid sequence; these have been termed subgroups I, II, and III (Kabat et al. 1983). Nucleic acid sequences corresponding to each of these subgroups have since been reported. Genes from subgroups I and III have been isolated using the homologous murine V family probes J558 and S107 respectively (Rechavi et al. 1982, 1983). More recently, a subgroup II gene (VCE-1) has been isolated as a J_H-associated rearrangement from an EBV-transformed human cell line (Takahashi et al. 1984). Nevertheless, information concerning the diversity and organization of these human V_H families remains scarce. As a necessary prerequisite to studies of human V_H utilization, we have begun to characterize human V_H genes and their evolutionary relationship to murine V_H genes (Berman et al. 1986). For these studies, we have employed the panel of mouse V_H family probes described above to identify and isolate homologous human V_H genes. Southern blotting of *Eco*R1- and *Hin*dIII-digested human DNA demonstrated that these murine probes identify unique as well as overlapping sets of genes (Table 1). The J558 family (mouse subgroup II) defines a unique family in human DNA (human subgroup I) consisting of approximately seven cross-reacting *Hin*dIII fragments; in contrast, J558 hybridizes to at least 50–60 novel fragments in murine DNA (BRODEUR and RIBLET 1984). Thus, the J558 family (human subgroup I), the largest murine family, seems to have contracted in humans relative to mice. The S107, J606, 7183, and 441-4 probes (mouse subgroup III) all hybridize to a second family of V genes, human subgroup III; although S107 and J606 probes are cross-reactive with only a subset of this family. The human subgroup III family contains approximately 18–20 hybridizing *Hin*dIII fragments, and thus appears to be similar in size to the mouse V_HIII subgroup [S107, 7183, J606, and 441-4 together hybridize to about 28 bands (BRODEUR and RIBLET 1984)]. We have confirmed the relationships suggested from the blotting analyses by directly isolating dozens of V_H genes from a human library on the basis of their hybridization of the various murine V_H family probes. Consistent with the Southern blotting data, J558 identified a set of cloned V genes distinct from those which hybridized to the four mouse subgroup III probes. Hybridization analysis of two human V genes which have been sequenced and identified as V_HI and V_HIII genes (MENSINK et al. 1986) provided support for these conclusions. The human V_HI gene from this line hybridizes solely to J558, while the V_HIII gene hybridizes to S107, J606, 441-4, and 7183 probes. On Southern blotting of human DNA, this human V_HIII probe yields a hybridization pattern indistinguishable from that produced by the 7183 probe. However, the human V_HI probe hybridizes to fragments in addition to those identified by the J558 probe. Thus, human V_HI can be divided into two subsets, one of which is more highly homologous to the mouse V_H J558 family.

Subgroup I in mouse is composed of the Q52 and 36–60 V_H families. At stringency conditions which normally distinguish between V_H families (i.e., no hybridization below 70%–80% homology), no fragments which hybridized to Q52 or 36–60 were found in human. Under less stringent hybridization conditions, a Q52 probe identified a set of three to six fragments in human. However, under these conditions Q52 cross-hybridizes with members of human subgroup III. Therefore, the possible relationship of the 36–60 and Q52 families to human subgroup II currently remains unclear. Two human genes, VCE-1 (TAKAHASHI et al. 1984) and V_HIG-1 (KUDO et al. 1985), have been classified as human subgroup II genes based on their translated amino acid sequences. Both are less than 65% homologous to Q52. VCE-1 is also less than 65% homologous to 36–60, although V_HIG-1 does have borderline family homology (71%) to 36–60. However, the nucleic acid sequences for VCE-1 and V_HIG-1 are less than 70% (62%) homologous to each other, and therefore by definition they should not be classified as family members. This suggests that human subgroup II is perhaps composed of two subfamilies, one of which is more closely related to the 36–60 family. Precedent for this is provided by the murine S107, 7183, J606, and 441-4 families, which by amino acid sequences were originally classified as a single subgroup (III) (KABAT et al. 1983).

5.2 Size of the Human V_H Repertoire

The size of a given murine V gene family has been roughly estimated as the number of fragments which hybridize to a family-specific probe on Southern blots of *Eco*R1-digested DNA (CORY et al. 1981); the size of the total murine V_H gene repertoire was taken to be the sum of the number of fragments hybridizing to probes for each of the different families. These estimates, ranging from 100 to 200 V_H genes, represent a minimum (KEMP et al. 1981; BRODEUR and RIBLET 1984). The occurrence of coincident hybridizing bands on Southern blots can result in an underestimate of V gene numbers; we have found clear evidence of such overlapping (in size) but clearly independent restriction fragments in studies of the murine V_HJ558 family (BERMAN and ALT, unpublished). Furthermore, the finding of at least two new and distinct murine V_H families (WINTER et al. 1985) raises the possibility that additional, as yet uncharacterized, families may exist. In human DNA, the estimated number of distinct hybridizing fragments which we detected by Southern blotting analyses was approximately 30; significantly, this figure is five- to tenfold less than the number of V_H-positive phage clones isolated from a library representative of a single human genome (BERMAN et al. 1986). The discrepancy could, again, be due in part to coincident bands. For example, preliminary analyses of cloned V_H genes suggest the existence of multiple overlapping, but distinct human *Hin*dIII fragments which contain human subgroup III V genes. Based on a combination of Southern blotting and library screening techniques, we roughly estimate that human subgroup III consists of 150–200 genes, while the J558-homologous subset of subgroup I is smaller, containing 20–40 genes. Our analyses suggest that subgroup II may be somewhat smaller (Table 1).

In general, V_H gene families are thought to be organized as typical multigene families. Accordingly, deletion mapping studies, studies of recombinant inbred strains, and analysis of phage and cosmid clones containing multiple V_H genes suggested that there is minimal interspersion of V_H families; that is, the various mouse V_H families, at least in BALB/c mice, appear to be segregated as homologous families along the chromosome (KEMP et al. 1981; BRODEUR and RIBLET 1984; BRODEUR et al. 1984; BOTHWELL 1984). Our studies suggest that a similar organization may not pertain to all human V_H gene families. Thus, we have determined that the inserts (average size 15 kb) of multiple independent recombinant phages contained both a V_HI and a V_HIII gene or a V_HIII and a V_HII gene (BERMAN et al. 1986). Similarly, others have recently shown that human V_κ gene families are interspersed (PECH and ZACHAU 1984). If intermingling proves to be a general phenomenon with respect to the organization of human V_H genes, it will make studies of position-related human V_H usage analogous to those described for mouse rather complex. In terms of V_H gene evolution, intermingling of V genes must be considered theories on generation of V gene diversity via amplification, contraction, and conversion mechanisms. For example, transposition-like events have been postulated in the evolution of human V_κ genes (PECH and ZACHAU 1984). Widespread interspersion of divergent V_H genes would also have significant implications with respect to theories which suggest a functional significance for clustering (and ordering) of particular V gene segments.

Acknowledgements. This work was supported by NIH grants AI-20047, CA-40427, and CA-2112-09 and by a Searle Scholars Award. BAM is supported in part by a Merck, Sharpe and Dohme fellowship and by a Leukemia Society of America fellowship. JEB is supported by a Damon Runyon–Walter Winchell Cancer Fund Fellowship, DRG-815. GDY is supported by NIH training grant GM-07367. FWA is an Irma T. Hirschl Career Scientist and a Mallinckrodt Scholar.

References

Alt FW (1986) New mechanism revealed. Nature 322:772

Alt FW, Rosenberg N, Enea V, Siden E, Baltimore D (1982) Multiple immunoglobulin heavy-chain gene transcripts in Abelson murine leukemia virus-transformed lymphoid cell lines. Mol Cell Biol 2:386

Alt FW, Yancopoulos GD, Blackwell TK, Wood C, Thomas E, Boss M, Coffman R, Rosenberg N, Tonegawa S, Baltimore D (1984) Ordered rearrangement of immunoglobulin heavy chain variable region segments. EMBO J 3:1209

Alt FW, Blackwell TK, DePinho RA, Reth MG, Yancopoulos GD (1986) Regulation of genome rearrangement events during lymphocyte differentiation. Immunol Rev 89:5

Baltimore D, Rosenberg N, Witte ON (1979) Transformation of immature lymphoid cells by A-MuLV. Immunol Rev 48:3

Berek C, Griffiths GM, Milstein C (1985) Molecular events during maturation of the immune response to oxazolone. Nature 316:412

Berman JE, Kowal C, Suh H, Mellis S, Alt FW (1986) Diversity and organization of human immunoglobulin V_H genes. Submitted

Blackwell TK, Yancopoulos GD, Alt FW (1984) Joining of immunoglobulin heavy chain variable region gene segments in vivo and within a recombination substrate. UCLA Symp Mol Cell Biol [NS] 19:537

Bosma GC, Custer RP, Bosma MJ (1983) A severe combined immunodeficiency mutation in the mouse. Nature 301:527

Bothwell ALM (1984) The genes encoding anti-NP antibodies in inbred strains of mice. In: Greene MI, Nisonoff A (eds) The biology of idiotypes. Plenum, New York, pp 19–34

Bothwell ALM, Paskind M, Reth M, Imanishi-Kari T, Rajewsky K, Baltimore D (1981) Heavy chain variable region contribution to the NP^b family of antibodies: somatic mutation evident in a 2a variable region. Cell 24:625

Brodeur PH, Riblet R (1984) The immunoglobIulin heavy chain variable region (IgG-V) locus in the mouse. I. One hundred IgH-V genes comprise seven families of homologous genes. Eur J Immunol 14:922

Brodeur PH, Thompson MA, Riblet R (1984) The content and organization of mouse IgH-V families. UCLA Symp Mol Cell Biol [NS] 18:445–453

Cancro MP, Wylie DE, Gerhard W, Klinman NR (1979) Patterned acquisition of the antibody repertoire: diversity of the hemaglutinin-specific B-cell repertoire in neonatal BALB/c mice. Proc Natl Acad Sci USA 76:6577

Cory S, Tyler BM, Adams JM (1981) Sets of Ig V_κ genes homologous to ten cloned V_κ sequences: implications for the number of germline V_κ genes. J Mol Appl Genet 1:103

Crews S, Griffin J, Huang H, Calame K, Hood L (1981) A single V_H gene segment encodes the immune response to phosphorylcholine: somatic mutation is correlated with the class of antibody. Cell 25:59

Denis KA, Klinman NR (1983) The genetic and temporal control of neonatal antibody expression. J Exp Med 157:1170

Dildrop R (1984) A new classification of mouse V_H sequences. Immunol Today 5:85

Dildrop R, Krawinkel V, Winter E, Rajewsky K (1985) V_H-gene expression in murine lipopolysaccharide blasts distributes over the nine known V_H-gene groups and may be random. Eur J Immunol 15:1154

Eisen HN, Reilly EB (1985) Lambda chains and genes in inbred mice. Annu Rev Immunol 3:337

Froscher BG, Klinman NR (1985) Strain-specific silencing of a predominant antidextran clonotype family. J Exp Med 162:1620

Gearhart PJ, Johnson ND, Douglas R, Hood L (1981) IgG antibodies to phosphorylcholine exhibit more diversity than their IgM counterparts. Nature 291:29

Hayakawa K, Hardy RR, Parks DR, Herzenberg LA (1983) The Ly-1 B cell subpopulation in normal, immunodefective and autoimmune mice. J Exp Med 157:202

Hayakawa K, Hardy RR, Honda M, Herzenberg LA, Steinberg AD, Herzenberg LA (1984) Ly-1 B cells: functionally distinct lymphocytes that secrete IgM autoantibodies. Proc Natl Acad Sci USA 81:2494

Kabat EA, Wu TT, Bilofsky H, Reid-Miller M, Perry H (1983) In: Sequences of proteins of immunoglobulin interest. NIH, Bethesda

Kemp DJ, Tyler B, Bernard O, Gough N, Gerondakis S, Adams JM, Cory S (1981) Organization of genes and spacers within the mouse immunoglobulin V_H locus. J Mol Appl Genet 1:245

Kleinfield R, Hardy RR, Tarlinton D, Dangl J, Henzenberg LA, Weigert M (1986) Recombination between an expressed immunoglobulin heavy chain gene and a variable gene segment in an Ly-1 B cell lymphoma. Nature 322:843–846

Klinman NR, Press JL (1975) The expression of specific clones during B cell development. Fed Proc 34:47

Kofler R, Perlmutter RM, Noonan DJ, Dixon FJ, Theofilopoulos AN (1985) Ig heavy chain variable region gene complex of lupus mice exhibits normal restriction fragment length polymorphism. J Exp Med 162:346

Kudo A, Ishihara T, Nishimura Y, Watanabe T (1985) A cloned human immunoglobulin heavy chain gene with a novel direct repeat sequence in 5′ flanking region. Gene 33:181

Lennon GG, Perry RP (1985) C_μ-containing transcripts initiate heterogeneously within the IgH enhancer region and contain a novel 5′-nontranslatable exon. Nature 318:475

Lewis S, Rosenbert M, Alt F, Baltimore D (1982) Continuing kappa-gene rearrangement in a cell line transformed by Abelson murine leukemia virus. Cell 30:807

Loh DY, Bothwell ALM, White-Scharf ME, Imanishi-Kari T, Baltimore D (1983) Molecular basis of a mouse strain-specific anti-hapten response. Cell 32:85

Lydyard PM, Grossi CE, Cooper MD (1976) Ontogeny of B cells in the chicken. I. Sequential development of clonal diversity in the bursa. J Exp Med 144:79

Manheimer-Lory AJ, Monestier M, Bellon B, Alt FW, Bona CA (1986) Fine specificity, idiotypy and nature of cloned V_H genes of murine monoclonal rheumaoid factor antibodies. Proc Natl Acad Sci USA 83:8293

Mensink EJBM, Schuurman RKB, Schot JDL, Thompson A, Alt FW (1986) Immunoglobulin heavy chain gene rearrangements in X-linked agammaglobulinemia. Eur J Immunol 16:963

Monestier M, Manheimer-Lory A, Bellon B, Painter C, Dang H, Talal N, Zanetti M et al. (1986) Shared idiotypes and restricted V_H genes characterize murine autoantibodies of various specificities. J Clin Invest 78:753

Nelson KJ, Kelley DE, Perry RP (1985) Inducible transcription of the unrearranged C_κ locus is a common feature of pre-B cells and does not require DNA or protein synthesis. Proc Natl Acad Sci USA 82:5305

Paige CJ (1983) Surface immunoglobulin-negative B-cell precursors detected by the formation of antibody-secreting colonies in agar. Nature 302:711–713

Palacios R, Henson G, Steinmetz M, McKearn JP (1984) Interleukin-3 supports growth of mouse pre-B cell clones in vitro. Nature 309:126

Pech M, Zachau HG (1984) Immunoglobulin genes of different subgroups are interdigitated within the V_κ locus. Nucleic Acids Res 12:9229

Perlmutter RM, Kearney JF, Chang SP, Hood LE (1985) Developmentally controlled expression of immunoglobulin V_H genes. Science 227:1597

Picard D, Schaffner W (1984) Unrearranged Ig variable region is transcribed in k-producing myelomas. EMBO J 3:3031

Press JL, Klinman NR (1974) Frequency of hapten-specific B cells in neonatal and adult mouse spleens. Eur J Immunol 4:155

Rechavi G, Bienz B, Ram D, Ben-Neriah Y, Cohen JB, Zakut R, Givol D (1982) Organization and evolution of immunoglobulin V_H gene subgroups. Proc Natl Acad Sci USA 79:4405

Rechavi G, Ram D, Glazer L, Zakut R, Givol D (1983) Evolutionary aspects of immunoglobulin heavy chain variable region V_H gene subgroups. Proc Natl Acad Sci USA 80:855

Reth MG, Alt FW (1984) Novel immunoglobulin heavy chains are produced from DJ_H gene segment rearrangements in lymphoid cells. Nature 312:418

Reth MG, Ammirati P, Jackson S, Alt FW (1985) Regulated progression of a cultured pre-B-cell line to the B-cell stage. Nature 317:353

Reth MG, Jackson S, Alt FW (1986a) V_HDJ_H formation and DJ_H replacement during pre-B differentiation: non-random usage of gene segments. EMBO J 5:2131–2138

Reth M, Gehrmann P, Petrac E, Wiese P (1986b) A novel V_H to V_HDJ_H joining mechanism in IgH-negative (null) pre-B cells results in heavy chain production. Nature 322:840–842

Riley SC, Connors SJ, Klinman NR, Ogata RT (1985) V gene segments that encode predominant DNP-specific clonotypes in neonatal BALB/c mice. In: Streilein JW, et al. (eds) Molecular biology of the immune system. Cambridge University Press, New York, pp 277–267 (Advances in gene technology, vol 2)

Selsing E, Storb U (1981) Mapping of immunoglobulin variable region genes: relationship to the "deletion" model of immunoglobulin gene rearrangement. Nucleic Acids Res 9:5725

Sherwin W, Rowlands D (1975) Determinants of the hierarchy of humoral immune responsiveness during ontogeny. J Immunol 115:1549

Sigal NH (1977) The frequency of *p*-azophenylarsonate and dimethylaminonapthalene-sulfonyl-specific B cells in neonatal and adult BALB/c mice. J Immunol 119:1129

Sigal NH, Gearhart PJ, Press JL, Klinman NR (1976) Late acquisition of a germ line antibody specificity. Nature 259:51

Silverstein A, Uhr J, Kramer K, Lukes R (1963) Fetal response to antigenic stimulus. II. Antibody production by the fetal lamb. J Exp Med 117:799

Spalding DM, Griffin JA (1986) Different pathways of differentiation of pre-B cell lines are induced by dendritic cells and T cells from different lymphoid tissues. Cell 44:507

Stohrer R, Kearney JF (1984) Ontogeny of B cell precursors responding to 1–3 dextran in BALB/c mice. J Immunol 133:2323

Takahashi N, Noma T, Honjo T (1984) Rearranged immunoglobulin heavy chain variable region V_H pseudogene that deletes the second complementarity-determining region. Proc Natl Acad Sci USA 81:5194

Teale JM (1985) B cell repertoire diversification in vitro. In: Streilein JW, et al. (eds) Molecular biology of the immune system. Cambridge University Press, New York, pp 319–320 (Advances in gene technology, vol 2)

Tonegawa S (1983) Somatic generation of antibody diversity. Nature 302:575

Ucker DS, Kurasawa Y, Tonegawa S (1985) An unusual clonotypic determinant on a cytotoxic T lymphocyte line is encoded by an immunoglobulin heavy chain variable region gene. J Immunol 135:4204

Van Ness BG, Weigert M, Coleclough C, Mather EL, Kelley DE, Perry RP (1981) Transcription of the unrearranged mouse C_κ locus: sequence of the initiation region and comparison of activity with a rearranged V_κ-C_κ gene. Cell 27:593

Whitlock CA, Witte ON (1982) Long term culture of B lymphocytes and their precursors from murine bone marrow. Proc Natl Acad Sci USA 79:3609

Winter E, Radbruch A, Krawinkel U (1985) Members of novel V_H gene families are found in VDJ regions of polyclonally activated B-lymphocytes. EMBO J 4:2861

Wood C, Tonegawa S (1983) Diversity and joining segments of mouse Ig H genes are closely linked and in the same orientation. Proc Natl Acad Sci USA 80:3030

Yancopoulos GD, Alt FW (1985) Developmentally controlled and tissue-specific expression of unrearranged V_H gene segments. Cell 40:271

Yancopoulos GD, Alt FW (1986) The regulation of variable region gene assembly. Annu Rev Immunol 4:339

Yancopoulos GD, Desiderio SV, Paskind M, Kearney JF, Baltimore D, Alt FW (1984) Preferential utilization of the most J_H-proximal V_H gene segments in pre-B-cell lines. Nature 311:727

Yancopoulos GD, Blackwell TK, Suh H, Hood L, Alt FW (1986a) Introduced cell receptor variable region gene segments recombine in pre-B cells: evidence that B and T cells use a common recombinase. Cell 44:251

Yancopoulos GD, DePinho RA, Lutzker SG, Zimmerman KA, Rosenberg N, Alt FW (1986b) Secondary genomic rearrangement events in pre-B cells: V_H DJ_H replacement by a LINE-/sequence and directed class-switching. EMBO J 5:3259

Yancopoulos GD, Bosma MJ, Alt FW (1986c) Regulation of V_H gene segment utilization during B-cell ontogeny. Submitted

Programmed Development of the Antibody Repertoire

R.M. Perlmutter

1 Introduction

Specific immunity in vertebrates requires the generation of a diverse repertoire of antigen-binding structures that are displayed on the surfaces of lymphocytes. These antigen receptors are encoded by a superfamily of gene segments which undergo rearrangements exclusively in lymphoid cells: the V_H, D_H, and J_H segments of antibody heavy chains, V_L and J_L segments of antibody light chains, and analogous components encoding the α, β, and γ-chains of the T-cell receptor (Hayday et al. 1985; Honjo 1983; Davis 1985; Kronenberg et al. 1986). In each case (with the possible exception of the T-cell receptor γ-chain genes), a limited repertoire of germ-line gene segments can be used to generate a much larger repertoire of recognition structures through combinatorial joining of these elements prior to transcription and through combinatorial association of the mature polypeptide products prior to expression of the heterodimeric antigen receptor on the cell surface.

Departments of Medicine (Medical Genetics) and Biochemistry and the Howard Hughes Medical Institute, SL-15, University of Washington, Seattle, WA 98195, USA

Current Topics in Microbiology and Immunology, Vol. 135

The mechanics of repertoire diversification have been most thoroughly investigated in the antibody genes. Studies of V_H, D_H, and J_H gene segment usage in hybridomas derived from LPS-stimulated adult splenocytes strongly suggest that the adult antibody repertoire is assembled through random combinatorial joining events (MANSER et al. 1984). At the same time, it is clear that the pattern of acquisition of responsiveness to a variety of antigens adheres to a developmental program. Thus fetal lymphocytes are able to generate antibodies to certain bacteriophage antigens, while responses to most protein antigens cannot be induced until late in the first week of extra uterine life (SHERWIN and ROWLANDS 1974; SILVERSTEIN et al. 1963). Developmentally ordered acquisition of antibody specificities is paradoxical if we assume that antibody gene segments rearrange randomly.

In the following pages I briefly review data suggesting that developmentally ordered rearrangements of antibody V_H gene segments in part underlie the programmed acquisition of antibody specificities. Though the data are imperfect and the conclusions necessarily tentative, they nevertheless permit development of a model for B-lineage stem cell maturation in which the pattern of V_H gene segment rearrangement depends on incremental changes in chromatin structure. This model has interesting implications regarding mechanisms of tolerance and immunodeficiency.

2 Development of the Antibody Repertoire

2.1 Mechanisms of Antibody Diversification

In the mouse, where the germ-line repertoire probably includes fewer than 1000 elements, combinatorial joining of V_H, D, and J_H gene segments (for heavy chains) and of V_L and J_L gene segments (for light chains) coupled with combinatorial association of the resulting heavy and light polypeptides can be calculated to yield more than 10^7 different antibody molecules (PERLMUTTER et al. 1984b). Direct examination of the products of gene rearrangement in adult antibody-forming cells supports a stochastic model for the generation of the antibody repertoire: single V_H regions appear associated with a random assortment of D_H, J_H, and light-chain elements (MANSER et al. 1984). The extent to which particular J or D gene segments are individually represented in the adult repertoire is still under investigation.

Superimposed on the already substantial diversity that results from combinatorial mechanisms is a process of localized somatic hypermutation which introduces small deletions, insertions, and single-base substitutions in and around rearranged antibody genes (TONEGAWA 1983; KIM et al. 1981). Somatic mutation is believed to be coupled to lymphocyte stimulation, perhaps through replication, and generates an even larger repertoire of antibody variable regions which are then positively selected by antigen (CLARKE et al. 1985; MANSER et al. 1984; RODWELL et al. 1983). In well-studied responses to phosphorylcholine (PC; PERLMUTTER et al. 1984b), azobenzenearsonate (ARS; SIEKEVITZ et al. 1983), group A streptococcal carbohydrate (GAC; PERLMUTTER et al. 1984a), influenza hemagglutinin (HA; CLARKE et al. 1985) and oxazolone (OX; KAARTINEN et al. 1983), somatic mutation exerts a dominant influence on the variability of natural antibody populations. Detailed analyses of mutations in the PC system suggest that these are introduced randomly, although perhaps in groups (PERLMUTTER et al. 1984b; GERHART and BOGENHAGEN 1983). Thus antibody diversity appears to result from mechanisms which act stochastically.

One potentially important contributor to the process of repertoire diversification is the structural organization of variable region gene segments which must rearrange. This organization has been most thoroughly investigated in the case of antibody V_H gene segments which altogether span perhaps 3 cM (as much as 6000 kb) of germ-line DNA (WEIGERT and RIBLET 1978). Murine V_H gene segments share common structural features and can be grouped by relative homology into families containing as few as two (V_HX24) or more than SO (V_HJ558) elements (BRODEUR and RIBLET 1984; KOFLER et al. 1985; SCHIFF et al. 1985). The spacing distance between individual V_H elements varies widely from a few kilobase pairs in the V_HJ558 family (BOTHWELL 1984) to more than 50 kb in the S107 V_H family (PERLMUTTER et al. 1984b). Genetic studies suggest that V_H gene families are discretely localized and ordered: centromere – V_HJ558–V_HS107–V_HQ52–V_H7183–D–J_H–C_μ on murine chromosome 12 (BRODEUR et al. 1984). This linkage of V_H gene segments within discrete homology groups may be maintained through frequent gene duplication and gene conversion events (PERLMUTTER et al. 1985a; PERLMUTTER and SCHROEDER 1986).

Considerably less is known regarding the organization of V_κ, V_α, and V_β gene segments. In each of these cases, however, individual variable region probes detect small families of closely homologous elements on genomic Southern blots (CORY et al. 1981; KRONENBERG et al. 1986).

2.2 Antibody Gene Rearrangement Is Developmentally Ordered

Although antibody diversity results largely from random combinatorial and mutational mechanisms as described above, detailed analysis of normal and transformed murine lymphocytes indicates that antibody gene rearrangements are developmentally regulated. B-lineage cells with self-renewing capability appear in murine fetal liver at about day 12 of gestation (OWEN et al. 1975; KINCADE 1981). The majority of identifiable cells of B-lymphoid lineage in fetal liver do not express surface antibody and are pre-B cells. The earliest rearrangements in such cells are D_H to J_H joining events, followed by V_H rearrangements and subsequently by light-chain rearrangements (MAKI et al. 1980; ALT et al. 1984). A similar pattern is seen in immature thymocytes, where γ-chain rearrangement precedes β and α rearrangement and D_β gene segment rearrangements precede V_β rearrangements (SAMELSON et al. 1985; SNODGRASS et al. 1985). In the adult, maturation of B lymphocytes within the bone marrow appears to recapitulate, at least in part, the developmental scheme outlined for fetal lymphoid cells as shown using Abelson-virus-transformed, bone-marrow-derived pre-B cell lines (YANCOPOULOS and ALT 1986). The fidelity of this rearrangement process is perhaps surprising in that the recombination signals located 3′ to variable-region gene segments, 5′ and 3′ to *D* gene segments, and 5′ to *J* gene segments are essentially identical for all members of the antibody gene superfamily (Kronenberg et al. 1986). More importantly, Abelson-virus-transformed pre-B cell lines will satisfactorily rearrange T-cell receptor β-chain gene constructs (Yancopoulos et al. 1986) as well as heavy- and light-chain gene constructs (Blackwell and Alt 1984; Lewis et al. 1985). Since normal pre-B cells do not typically rearrange T-cell receptor gene segments, some other regulatory mechanism must define which substrates will be active in rearrangement (Yancopoulos et al. 1986).

2.3 Feedback Regulation of Antibody Gene Rearrangement

Some insight into the process of rearrangement has been gained from studies of mice made transgenic for a productively rearranged antibody heavy chain (Grosschedl et al. 1984; Weaver et al. 1985) or light chain (Ritchie et al. 1984). In these animals, the presence of a functional antibody protein appears to limit rearrangements of endogenous antibody genes in B-lymphocytes, thus suggesting that antibody gene rearrangements are regulated by feedback mechanisms. According to the feedback model, the B-cell recombinase is inititially predisposed to rearrange D_H gene segments. Subsequent satisfactory rearrangement of a V_H gene segment results in production of a heavy-chain protein product which simultaneously signals an end to further V_H–D_H rearrangement and initiates V_L–J_L joining events (Reth et al. 1985). Light-chain rearrangements continue, first at the κ locus and then at the λ locus, until a satisfactory antibody protein is formed. In this way, each individual B cell synthesizes products of only a single heavy- or light-chain allele. Feedback inhibition of antibody rearrangement is clearly not absolute, since the presence of productively rearranged transgenic antibody genes does not preclude all endogenous rearrangements (Rusconi and Kohler 1985). Alternative views of antibody gene rearrangement that do not invoke feedback regulation at the level of antibody protein are of course still tenable (Coleclough 1983).

2.4 The Acquisition of Antibody Specificities Is Developmentally Programmed

Results of fetal and neonatal immunization in sheep (Silverstein et al. 1963), mice (Sherwin and Rowlands 1974), and in man (Smith et al. 1964) first suggested that responsiveness to particular antigens appears at characteristic times during mammalian development. This phenomenon has since been studied in greatest detail in the mouse using adoptive transfer methods and most particularly the splenic focus assay of Klinman and collaborators (Klinman and Press 1975). Antibody-forming cell precursors reactive with some antigens, notably dinitrophenyl (DNP) haptens, are detectable by day 14 of fetal life, while precursors reactive with many other antigens, e.g., PC (FUNG and KOHLER 1980; SIGAL et al. 1977) or α-(1→3) dextran (STOHRER and KEARNEY 1984), do not appear until late in the first neonatal week.

One theory which might explain the developmentally ordered appearance of antibody specificities posits that antibody V_H and V_L gene segments rearrange randomly at all times during development, but some superimposed regulatory mechanism, perhaps focusing on the idiotype of the early surface immunoglobulin, selects which clones are amplified. This external regulatory model seems unlikely for two reasons. First, the early absence of B cells reactive with inulin (BONA et al. 1979) and α-(1→6) dextran (HOWARD and HALE 1976) reflects an absolute lack of antigen-specific B-cell precursors (although this could be argued to result from a selection mechanism specifically deleting cells bearing the appropriate surface antibodies). Secondly, antibody specificities are acquired in a consecutive way in fetal liver fragments explanted in vitro and hence removed from constraints imposed by other organs within the developing embryo (TEALE 1985). While this does not preclude an influence of adventitial cells situated adjacent to developing pre-B cells within the hepatic parenchyma, it is unlikely that these cells would have the specificity necessary to distinguish amongst different antibody specificities. Thus it seems likely that the developmental program governing the acquisition of antibody specificities is intrinsic to B-lineage cells themselves.

To help understand the process of antibody repertoire diversification, a variety of studies have sought to examine the rearrangement of antibody V gene segments at different times during development.

2.5 Developmentally Programmed Rearrangement of V_H Gene Segments

Since antibody V_H gene segments can be conveniently grouped into non-cross-hybridizing families, it is possible to examine the developmental program of V_H family rearrangement in mRNA derived from fetal and adult B-lineage cells by Northern blot hybridization. When this is done using pooled splenic or fetal liver mononuclear cells, however, the potential for artifact is great, since contamination with just a few mature B lymphocytes expressing high levels of antibody mRNA will skew the results greatly. Thus experiments have focused on early and late B-lineage clones derived using Abelson-virus-transformed cell lines (ALT et al. 1981) or fetal and neonatal liver hybridomas (BURROUGHS et al. 1979).

YANCOPOULOS et al. (1984) examined nine Abelson-virus-transformed cell lines, generated through infection of BALB/c liver, and documented 16 rearrangements in these cells, of which 15 involved a member of the V_H7183 family. Strikingly, in nine cases the same V_H gene segment, V_H81X, was rearranged. Similar results were obtained by GEARHART and LIN (1985), who examined seven 14- to 17-day BALB/c fetal pre-B cell lines and found six had rearranged V_H81X. This preference for V_H7183 gene family rearrangements is not a feature of adult BALB/c B lymphocytes, since only eight of 21 characterized rearrangements in marrow-derived cell lines used a member of the V_H7183 family (YANCO-

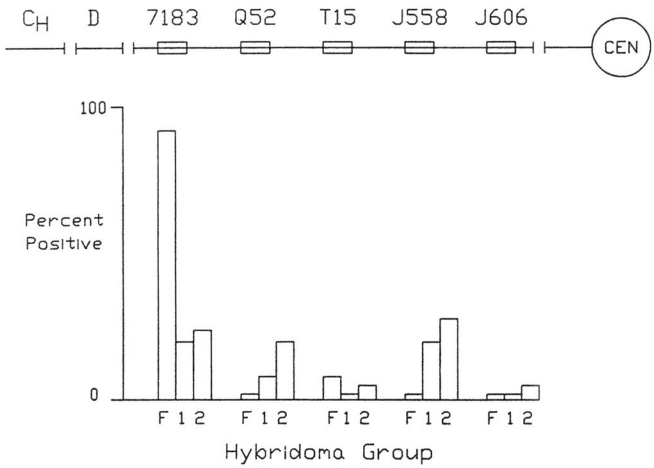

Fig. 1. The chromosomal order of V_H gene families may dictate the order of V_H gene expression during development. The architecture of the murine V_H locus is diagrammed at the *top* and compared with the timing of expression of the indicated V_H gene families in fetal (*F*), neonatal day 1 (*1*), and neonatal day 2 (*2*) liver hybridomas. (Adapted from Perlmutter et al. 1986)

POULOS et al. 1984). Thus fetal B-lineage cells utilize a restricted subset of those V_H gene segments available to adult lymphocytes.

Our own studies of early B and pre-B cell hybridomas have substantially extended the results obtained using Abelson virus transformants. As shown graphically in Fig. 1, of 15 fetal liver hybridomas, 11 expressed a member of the V_H7183 family (PERLMUTTER et al. 1985b). In contrast, only 20% of day-1 and day-2 neonatal hybridomas used a member of this V_H gene family, while expression of other V_H gene families increased progressively, especially for the V_HJ558 and V_HQ52 families (Fig. 1). High expression of the V_H7183 family in fetal pre-B cells (78%) and the lesser representation of this family in neonatal pre-B cells (0%) cannot reflect selection by antigen or by idotype-specific regulatory cells, since these pre-B cells do not synthesize light chain and hence cannot assemble and express antibody molecules (PERLMUTTER et al. 1985b). It thus appears that as in the case of the generation of antibody specificities, V_H gene expression follows a developmental program generated within B-lineage cells or in the stem cells from which they are derived.

2.6 Molecular Architecture of the V_H Locus and the Timing of V_H Gene Rearrangement

One mechanism which may possibly underlie the developmental program of V_H gene rearrangement is suggested by examination of the large-scale organization of the V_H locus. In Fig. 1, the chromosomal organization of five murine V_H gene families is superimposed onto a tabulation of the timing of expression of these families, depicted in bar-graph format, in fetal and neonatal hybridomas. Remarkably, the chromosomal order of the D_H and V_H elements roughly correlates with the time during development when each is expressed. Thus rearrangements begin in the D_H region located proximal to the C_μ gene and progress centromerically to include the V_H7183 family in late fetal life. Rearrangements of the most centromeric V_H family, J606, do not appear until the 2nd day of neonatal life. The relatively large number of V_HJ558 rearrangements compared with other more telomeric gene families may reflect the greater size of this family (>50 elements: BOTHWELL 1984; BRODEUR and RIBLET 1984; KOFLER et al. 1985; SCHIFF et al. 1985). Family size cannot, however, explain the differences in the timing of rearrangement of the V_H7183 and V_HJ606 families, since each of these contains between 10 and 12 cross-hybridizing species (BRODEUR and RIBLET 1984; KOFLER et al. 1985). These data raise the possibility that the developmental program of heavy-chain rearrangement is inscribed linearly on murine chromosome 12.

How might chromosomal order and the timing of gene rearrangement be correlated? Since the enzymatic machinery responsible for rearrangement appears able to catalyze recombinations between gene segments possessing only a few hundred base pairs of antibody gene sequences (YANCOPOULOS et al. 1986), regulation of V_H gene segment rearrangement probably depends on factors which limit access of these putative recombinases to particular chromosomal regions. Antibody gene constructs which are substrates for rearrangement in B-lineage cells are typically found within DNAse-sensitive, "open" chromatin structures (YANCOPOULOS et al. 1986). In contrast, the nonrearranging endogenous T-cell receptor β genes in the same cells are relatively DNAse-resistant.

One measure of the increased recombinase accessibility of V_H regions poised for rearrangement may be transcriptional activation at the rearranging locus. Synthesis of atypical transcripts from germ-line elements clearly precedes rearrangement in the J_κ locus (NELSON et al. 1985), in T-cell receptor β-chain genes (KRONENBERG et al. 1986), and in the C_μ locus as well (RETH and ALT 1984). More importantly, transcription of unrearranged V_H gene segments appears to precede V_H to D-J_H joining in pre-B cells but does not occur in mature B-lymphocytes (YANCOPOULOS and ALT 1985). With this admittedly preliminary information in mind, the developmentally ordered expression of antibody V_H genes can be viewed as the progressive activation, from telomere to centromere, of the V_H locus.

2.7 Maturation of B-Lineage Stem Cells and the Program of V_H Gene Expression

B-lymphocyte precursors divide quite actively in sites of lymphopoiesis. For example, the average production of lymphocytes exceeds 10^8/h in lamb Peyer's patches (REYNOLDS and MORRIS 1984). The stem cells from which these B-lymphocytes derive must contain germ-line antibody genes. Thus the rearrangement machinery, activated in pre-B cells, will interact with a substrate chromatin configuration inherited from the stem cell population. Figure 2 presents a model of B-lineage stem cell maturation in which the antibody V_H locus becomes pro-

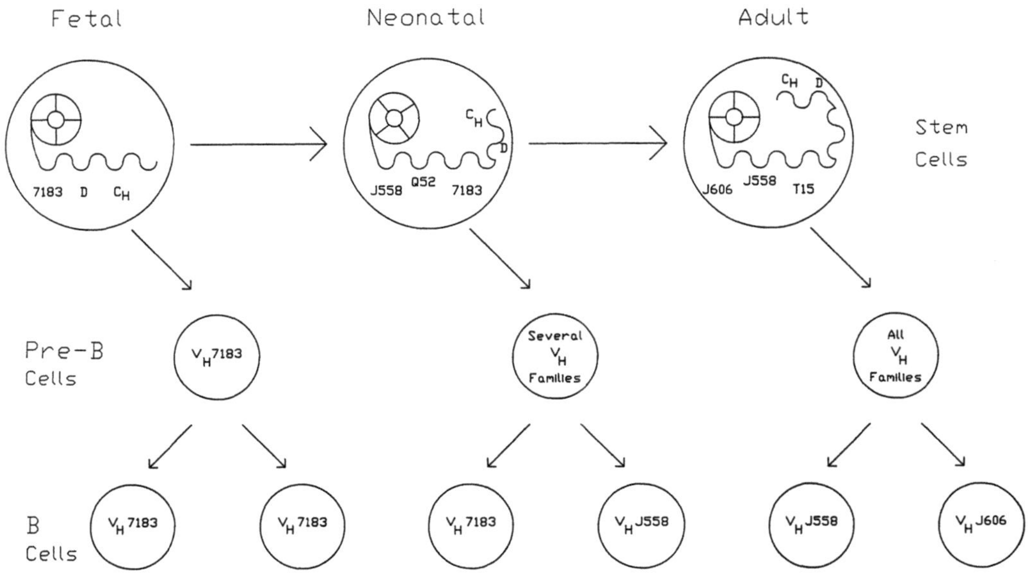

Fig. 2. A model for progressive alterations in chromatin structure associated with the maturation of B-lineage stem cells. Increases in chromatin accessibility are denoted by the unwinding of a metaphoric tape spool in the maturing stem cell population. The specific V_H families available for expression are indicated. (Reproduced from Perlmutter et al. 1986)

gressively more accessible to recombinase. Early in development, the heavy-chain locus may be tightly bound in higher-order structures that do not permit recombinase access. As development proceeds, these higher-order structures relax and unfold in a characteristic fashion, maturing in a centromeric direction. This process is depicted in Fig. 2 as the unreeling of a chromatin tape spool. Stem cells at each developmental timepoint divide to yield daughter cells that may rearrange only those V_H gene segments located in an open chromatin region. Fetal cells might thus be able to rearrange only those V_H gene segments located at the telomeric end of the heavy chain locus. In adult B stem cells, the entire V_H region would fall within an open chromatin segment and thus rearrangements in the progeny of these cells could involve any combination of V_H, D_H, and J_H elements. This prediction is of course in accord with experimental data on the stochastic generation of the adult antibody repertoire (Manser et al. 1984).

Correlations between chromosomal order and the timing of expression of genes during development have been observed in other cell lineages. In hematopoiesis in man, for example, maturation of a single stem cell population (Stamatoyannopoulos et al. 1983) results in expression of ε-globin genes, followed by γ-globin genes and finally by mature adult β-globins, in precise accord with the transcriptional order of these elements on chromosome 11 (Efstratiadis et al. 1980). Similarly, the chromosomal organization of the *bithorax* genes in *Drosophila* parallels the anatomic order of the segments which these genes control and the temporal order of their expression (Lawrence and Morata 1983).

In each of these cases, it is possible that progressive alterations in chromatin structure permit the programmed expression of a family of related genetic elements.

3 Programmed Rearrangement of V_H Gene Segments and Immune Physiology

3.1 Does the Timing of V_H Gene Expression Explain the Developmentally Ordered Acquisition of Antibody Specificities?

Granted that antibody V_H gene segments rearrange with characteristic and variable frequencies at different times during development, is the pattern of V_H rearrangement sufficient to explain the programmed acquisition of the antibody repertoire? Clearly not. Table 1 lists a group of antigens for which the age at which antibody-forming cell precursors can be demonstrated is known. Juxtaposed with each such antigen is the principle V_H gene family that contributes to the response and the relative time (fetal, neonatal, adult) at which rearrangements are seen to involve each family. The time level of resolution for antibody V_H gene rearrangements is of course rather low, and thus the only assertion that can comfortably be made is that antigens which recruit late-rearranging V_H gene families should not appear at early developmental timepoints. As shown in Table 1, there are no exceptions to this rule.

Interestingly, antigens which stimulate responses in fetal B cells, such as DNP, elicit antibodies that frequently utilize members of the V_H7183 gene family (Riley et al. 1986) or of the V_H36–60 family (Teale and Kearney 1986), which has not yet been fully evaluated in terms of either its chromosomal location or the time at which its elements typically rearrange. Antibodies that utilize late-rearranging V_H elements (J558, J606) are frequently induced by antigens

Table 1. The order of acquisition of antigen responsiveness compared with the developmentally programmed rearrangement of antibody V_H gene families

Antigen	Time of appearance of reactive B cells	Relevant V_H gene family	Time of rearrangement
DNP[a]	Fetal day 14	7183, 36–60	Fetal, unknown
HA[b]	By neonatal day 3	7183, S107	Fetal, late fetal
PC[c]	First neonatal week	S107	Late fetal
α-(1→3) Dextran[d]	Neonatal day 5	J558	Neonatal
α-(1→6) Dextran[e]	Late neonatal	J606 (Q52)	Late neonatal (early neonatal)

Based on adoptive transfer studies and the data presented in Fig. 1. See text for details

[a] Klinman and Press (1975), Riley et al. (1986)

[b] Cancro et al. (1979), Clarke et al. (1985)

[c] Sigal et al. (1977), Perlmutter et al. (1984b)

[d] Stohrer and Kearney (1984), Hansburg et al. (1978)

[e] Bona et al. (1979), Perlmutter et al. (1984a)

that do not stimulate B-cell responses until after birth [α-(1→3) dextran, inulin]. These observations are, of course, in perfect accord with predictions based on the model for B-lineage stem cell maturation depicted in Fig. 2.

In some cases, antigens which stimulate responses relatively late in development utilize a V_H gene segment that may rearrange relatively early (for example, PC antigens and the S107 gene family). This phenomenon might be explained if V_κ gene rearrangements are also developmentally ordered and if the antigen recruits precursors which utilize a late-rearranging V_κ gene segment (PERLMUTTER et al. 1985b). By extrapolation it seems likely that variable regions encoding the T-cell antigen receptor may also undergo developmentally ordered rearrangements.

At the same time, a model which attributes maturation of the antibody repertoire to alterations in the frequency of variable region gene rearrangement cannot explain the recruitment of unexpected B-lineage precursors from neonates through specialized immunization protocols. For example, neonatal priming using anti-idiotypic reagents renders young animals responsive to otherwise nonimmunogenic bacterial polysaccharides (STEIN 1985). Thus other regulatory mechanisms can clearly operate to expand, and presumably suppress, limited pools of neonatal B lymphocytes bearing antibodies typically encoded by genes expressed at later stages of maturation. It is worth mentioning in this context that nearly 20% of fetal B lymphocytes express V_H genes from families other than V_H7183 (Fig. 1).

3.2 B-Cell Repertoire Dynamics and the Induction of Tolerance

The developmentally programmed expression of V_H gene families has interesting consequences for theories of acquisition of immunocompetence and of tolerance. Clonal deletion events based on the increased susceptibility to tolerance of immature B cells have been proposed to explain the absence of autoreactive antibodies in normal adults (CAMBIER et al. 1976). If, as I have suggested, the majority of the B-cell repertoire is generated after birth, tolerance cannot, in general, be developed to antigens seen in utero. Thus if neonatal tolerance reflects a specific suppression of B-cell specificities induced through interaction of antigen with cell-surface antibodies (SZEWCZUK and SISKIND 1977), the majority of these interactions will involve persistent immunogen that remains in a recognizable form for some days after administration. This type of reasoning is also at odds with theories of immune receptor diversification which rely on early "network" interactions between complementary antibody molecules as a means of repertoire assembly. The receptor repertoire will be limited in early life to a small subset of V_H regions selected on the basis of chromosomal position and evolutionary relatedness (homology), and not on the basis of complementarity.

3.3 Evolutionary Selection for Programmed Repertoire Diversification

Humoral immune defense is provided to fetal and early neonatal animals through maternal antibodies; thus, limitation of antibody gene rearrangements early in life probably will not result in immune compromise. If the germ-line antibody repertoire is prejudiced in favor of combining sites with antipathogen specificity (PERLMUTTER and SCHROEDER 1986), the developmental program of antibody gene expression may have been selected to permit early expression of highly protective antibody species. Alternatively, V_H rearrangements may be specifically limited in fetal organisms in order to minimize the development of immune responses to maternal or self antigens at a time when more complex regulatory mechanisms have not yet developed.

Genetic analysis provides a potential means of examining the importance of the developmental program of repertoire acquisition. While it is clear that most inbred strains contain an extremely similar complement of antibody V_H gene segments (KOFLER et al. 1985; PERLMUTTER et al. 1985a), there may well be some strain-specific variation in the architecture of the V_H locus and in the developmental pattern of V_H gene rearrangement. Virtually all of the analyses of V_H gene expression during development published to date have utilized BALB/c or C57BL cells (GEARHART and LIN 1985; PERLMUTTER et al. 1985b; YANCOPOULOS et al. 1984). Recent evidence indicates that the pattern of V_H restriction varies considerably in different inbred strains. In one set of fetal day-16 hybridomas generated using NZB mice, the V_H7183 gene family was completely unrepresented (S. Pawar, J.L. Klotz, and R.M. Perlmutter, unpublished data). This observation is especially interesting, since B-lineage precursors arise prematurely and in elevated numbers in NZB mice, reflecting the action of a humoral factor which promotes B-cell development (JYONOUCHI and KINCADE 1984). Detailed analysis of these strain differences will aid in testing whether the timing of V_H gene family rearrangements is indeed dependent upon the chromosomal order of these elements, and will enable a careful study of alterations in chromatin structure associated with V_H expression.

3.4 The Development of the Antibody Repertoire and Immunopathology

Specific immune abnormalities might be expected to result from alterations in the normal developmental program of V_H expression. For example, retardation of the normal maturation process would fix B-lineage stem cells in an early developmental configuration resulting in expression of an incomplete antibody repertoire. In man, X-linked agammaglobulinemia occurs with a frequency of about 1 in 10^5 live births (HOSKINE and ROBERTSON 1983). Individuals afflicted with this disorder comprise a spectrum including patients with essentially absent serum immunoglobulins or, more commonly, patients with demonstrable serum immunoglobulin but absent responses to a variety of model antigens (LEVITT et al. 1984; OCHS et al. 1971). These are precisely the phenotypes that would be expected to result from a variable block in the maturation of B-lineage precursors. In some affected individuals, Epstein–Barr-virus-transformed B-lymphoblastoid cell lines obtained from peripheral blood contain D_H–J_H rearrangement events and D_H–J_H transcripts without evidence of V_H rearrangement (SCHWABER et al. 1983). This may represent persistence of a closed chromatin structure throughout the V_H locus as represented in Fig. 2.

It is provocative that mice bearing the X-linked *xid* mutation which lack mature Lyb5$^+$ lymphocytes are in general unresponsive to polysaccharide antigens (reviewed in Scher 1982) which typically recruit late-rearranging V_H gene segments (Table 2). A defect in B-lineage stem cell maturation resulting in abnormal chromatin structure in the heavy-chain locus could in part explain this phenotype. At the same time, the existence of distinct B-cell subpopulations emphasizes the oversimplification inherent in the scheme depicted in Fig. 2: the developmental pattern of V_H gene expression may reflect alterations in precursor frequencies for multiple, semi-independent B-cell lineages rather than maturation of a single B-lineage stem cell population (Hardy et al. 1984).

4 Future Directions

While there is still much to be learned from analyses of the frequency of specific V_H rearrangements in pre-B cells derived from fetal and early neonatal animals, a thorough understanding of the maturation of the antibody repertoire will require experimental manipulation of B-cell differentiation. In this regard, bone marrow culture systems (Whitlock et al. 1985) will probably prove indispensable. Since development of the antibody repertoire appears to proceed normally in fetal liver fragments (Teale 1985), it is legitimate to ask whether progressive alterations in the frequency of V_H gene expression are observed in cultured fetal B-lineage cells and whether this maturation process can be accelerated or retarded by changes in the extracellular milieu. Such approaches will also permit direct evaluation of the recombinase accessibility model of rearrangements in the antibody gene superfamily (Perlmutter et al. 1986; Yancopoulos et al. 1986). Finally, especially useful information may well be derived from the application of these approaches to the study of human immunodeficiency diseases in which arrested development of the B-cell lineage appears to represent the primary immune defect (Perlmutter et al. 1986).

References

Alt FW, Rosenberg N, Lewis S, Thomas E, Baltimore D (1981) Organization and reorganization of immunoglobulin genes in A-MuLV-transformed cells: rearrangement of heavy but not light chain genes. Cell 27:381–390

Alt FW, Yancopoulos GC, Blackwell TK, Wood C, Thomas E, Boss M, Coffman R, Rosenberg N, Tonegawa S, Baltimore D (1984) Ordered rearrangement of immunoglobulin heavy chain variable region segments. EMBO J 3:1209–1219

Blackwell TK, Alt FW (1984) Site-specific recombination between Ig D and J_H segments that were introduced into the genome of a murine pre-B cell line. Cell 37:105–112

Bona C, Mond JJ, Stein KE, House S, Lieberman R, Paul WE (1979) Immune response to levan. III. The capacity to produce anti-inulin antibodies and cross-reactive idiotypes appears late in ontogeny. J Immunol 123:1484–1491

Bothwell A (1984) The genes encoding anti-NP antibodies in inbred strains of mice. In: Greene MI, Nisonoff A (eds) The biology of idiotypes. Plenum, New York pp 19–34

Brodeur PH, Riblet R (1984) The immunoglobulin heavy chain variable region (Igh-V) locus in the mouse. I. 100 Igh-V genes comprise 7 families of homologous genes. Eur J Immunol 14:922–930
Brodeur PH, Thompson MA, Riblet R (1984) The content and organization of mouse Igh-V families. UCLA Symp Mol Cell Biol [NS] 18:445–453
Burroughs P, LeJeune M, Kearney JF (1979) Evidence that murine pre-B cells synthesize μ heavy chain but no light chains. Nature 280:838–840
Cambier JC, Kettman JR, Vitetta ES, Uhr JW (1976) Differential susceptibility of neonatal and adult murine spleen cells to in vitro induction of B-cell tolerance. J Exp Med 144:293–301
Cancro MP, Wylie DE, Gerhard W, Klinman NR (1979) Patterned acquisition of the antibody repertoire: diversity of the hemagglutinin-specific B cell repertoire in neonatal BALB/c mice. Proc Natl Acad Sci. USA 76:6577–6581
Clarke SH, Huppi K, Ruezinsky D, Staudt L, Gerhard W, Weigert M (1985) Inter- and intraclonal diversity in the antibody response to influenza hemagglutinin. J Exp Med 161:687–704
Coleclough C (1983) Chance, necessity and antibody gene dynamics. Nature 303:23–25
Cory S, Tyler B, Adams J (1981) Sets of immunoglobulin V_κ genes homologous to 10 cloned V_κ sequences: implications for the number of germline V_κ genes. J Mol Appl Genet 1:103–116
Davis MM (1985) Molecular genetics of T-cell receptor beta chain. Annu Rev Immunol 3:537–560
Efstratiadis A, Posakony JW, Maniatis T, Lawn RM, O'Connell C, Spritz RA, DeRiel JK, et al. (1980) The structure and evolution of the human β-globin gene family. Cell 21:653–668
Fernandez C, Moller G (1978) Immunological unresponsiveness to native dextran B512 in young animals of dextran high responder strains is due to lack of Ig receptor expression. Evidence for a non-random expression of V-genes. J Exp Med 147:645–655
Fung JJ, Kohler H (1980) Late clonal selection and expansion of the TEPC-15 germline specificity. J Exp Med 153:1262–1270
Gearhart PJ, Bogenhagen D (1983) Clusters of point mutations are found exclusively around rearranged antibody variable region genes. Proc Natl Acad Sci. USA 80:3439–3443
Gearhart PJ, Lin P (1985) Development of the B cell repertoire. In: Streilein JW, Ahmed F, Black S, Blomberg B, Voellmy RW (eds) Molecular Biology of the Immune System. Cambridge University Press, Cambridge, pp 7–10 (Advances in gene technology)
Grosschedl R, Weaver D, Baltimore D, Costantini F (1984) Introduction of a μ immunoglobulin gene into the mouse germ line: specific expression in lymphoid cells and synthesis of functional antibody. Cell 38:647–658
Hansburg D, Perlmutter RM, Briles DE, Davie JM (1978) Analysis of the diversity of murine antibodies to dextran B1355. III. Idiotypic and spectrotypic correlations. Eur J Immunol 8:352–361
Hardy RR, Hayakawa K, Parks DR, Herzenberg LA, Herzenberg LA (1984) Murine B cell differentiation lineages. J Exp Med 159:1169–1191
Hayday AC, Saito H, Gillies SD, Kranz DM, Tanigawa G, Eisen HN, Tonegawa S (1985) Structure organization and somatic rearrangement of T-cell gamma genes. Cell 40:259–269
Honjo T (1983) Immunoglobulin genes. Annu Rev Immunol 1:499–528
Hoskine CS, Robertson DM (1983) Epidemiology and treatment of hypogammaglobulinemia. Birth Defects 19:223–227
Howard JG, Hale C (1976) Lack of neonatal susceptibility to induction of tolerance to polysaccharide antigens. Eur J Immunol 6: 486–496
Jyonouchi H, Kincade PW (1984) Precocious and enhanced functional maturation of B lineage cells in New Zealand Black mice during embryonic development. J Exp Med 159:1277–1282
Kaartinen M, Griffiths G, Markham AF, Milstein C (1983) mRNA sequences define an unusually restricted IgG response to 2-phenyloxazolone and its early diversification. Nature 304:320–325
Kim S, Davis MM, Sinn E, Patten P, Hood L (1981) Antibody diversity: Somatic hypermutation of rearranged V_H genes. Cell 27:573–581
Kincade PW (1981) Formation of B lymphocytes in fetal and adult life. Adv Immunol 31:177–245
Klinman NR, Press JL (1975) The B cell specificity repertoire: its relationship to definable subpopulations. Transplant Rev 24:41–83
Kofler R, Perlmutter RM, Noonan DJ, Dixon FJ, Theofilopoulos AN (1985) Ig heavy chain variable region gene complex of lupus mice exhibits normal restriction fragment length polymorphism. J Exp Med 162:346–351
Kronenberg M, Siu G, Hood L, Shastri N (1986) The molecular genetics of the T-cell antigen receptor and of T-cell antigen recognition. Annu Rev Immunol 4, 529–591

Lawrence PA, Morata G (1983) The elements of the bithorax complex. Cell 35:595–601
Levitt D, Ochs HD, Wedgwood RJ (1984) Epstein-Barr virus-induced lymphoblastoid cell lines derived from the peripheral blood of patients with X-linked agammaglobulinemia can secrete IgM. J Clin Immunol 4:143–150
Lewis S, Gifford A, Baltimore D (1985) DNA elements are asymmetrically joined during the site-specific recombination of kappa immunoglobulin genes. Science 228:677–685
Loh DY, Bothwell ALM, White-Scharf ME, Imanishi-Kari T, Baltimore D (1984) Molecular basis of a mouse strain-specific anti-hapten response. Cell 33:85–93
Maki R, Kearney J, Paige C, Tonegawa S (1980) Immunoglobulin gene rearrangements in immature B cells. Science 209:1366–1369
Manser T, Huang S-Y, Gefter M (1984) Influence of clonal selection on the expression of immunoglobulin variable region genes. Science 226:1283–1287
Nelson KJ, Kelley DE, Perry RP (1985) Inducible transcription of the unrearranged C_κ locus is a common feature of pre-B cells and does not require DNA or protein synthesis. Proc Natl Acad Sci. USA 82:5305–5309
Ochs HD, Davis SD, Wedgwood RJ (1971) Immunologic response to bacteriophage ϕX174 in immunodeficiency diseases. J Clin Invest 50:2559–2568
Owen JJT, Raff MC, Cooper MD (1975) Studies on the generation of B lymphocytes in the mouse embryo. Eur J Immunol 5:468–473
Perlmutter RM, Schroeder HW Jr (1987) Molecular evolution of the T15 V_H gene family. In: Kelsoe G (ed) Evolution and vertebrate immunity. University of Texas Press, Austin, in press
Perlmutter RM, Klotz JL, Bond MW, Nahm M, Davie JM, Hood L (1984a) Multiple V_H gene segments encode murine antistreptococcal antibodies. J Exp Med 159:179–192
Perlmutter RM, Crews ST, Douglas R, Sorensen G, Johnson N, Nivera N, Gearhart PJ, Hood L (1984b) The generation of diversity in phosphorylcholine-binding antibodies. Adv Immunol 35:1–37
Perlmutter RM, Berson B, Griffin JA, Hood L (1985a) Diversity in the germline antibody repertoire: molecular evolution of the T15 V_H gene family. J Exp Med 162:1998–2016
Perlmutter RM, Kearney JF, Chang SP, Hood LE (1985b) Developmentally controlled expression of immunoglobulin V_H genes. Science 227:1597–1601
Perlmutter RM, Pawar S, Schroeder HW Jr (1986) Programmed development of the antibody repertoire: immunopathologic implications. In: Eibl M, Cooper MD, Wedgdwood R (eds) Primary immunodeficiency diseases. Excerpta Medica, Amsterdam, pp 3–12
Reth MG, Alt FW (1984) Novel immunoglobulin heavy chains are produced from D–J_H gene segment rearrangements in lymphoid cells. Nature 312:418–423
Reth MG, Amirati P, Jackson S, Alt FW (1985) Regulated progression of a cultured pre-B cell line to the B cell stage. Nature 317:353–355
Reynolds JD, Morris B (1984) The emigration of lymphocytes from Peyer's patches in sheep. Eur J Immunol 14:7–13
Riley SR, Klinman NR, Ogata RJ (1986) Preferential expression of V_H gene segments by predominant DNP-specific BALB/c neonatal antibody clonotypes. Proc Natl Acad Sci. USA 83:2589–2593
Ritchie KA, Brinster RL, Storb U (1984) Allelic exclusion and control of endogenous immunoglobulin gene rearrangement in κ transgenic mice. Nature 312:517–520
Rodwell JD, Gearhart PJ, Karush F (1983) Restriction in IgM expression. IV. Affinity analysis of monoclonal anti-phosphorylcholine antibodies. J Immunol 130:313–316
Rusconi S, Kohler G (1985) Transmission and expression of a specific pair of rearranged immunoglobulin μ and κ genes in a transgenic mouse line. Nature 314:330–335
Samelson LE, Lindsten T, Fowlkes BJ, van den Eisen P, Terhorst C, Davis MM, Germain RN, Schwartz RH (1985) Expression of genes of the T cell antigen receptor complex in precursor thymocytes. Nature 315:765–768
Scher I (1982) CBA/N immune defective mice: evidence for the failure of a B cell subpopulation to be expressed. Immunol Rev 64:117–136
Schiff C, Milili M, Fougereau M (1985) Functional and pseudogenes are similarly organized and may equally contribute to the extensive antibody diversity of the IGV_HII family. EMBO J 4:1125–1230
Schwaber J, Molgaard H, Orkin S, Gould H, Rosen F (1983) Early pre-B cells from normal and X-linked agammaglobulinemia produce C_μ without associated variable region. Nature 302:355–358

Sherwin W, Rowlands DT Jr (1974) Development of humoral immunity in lethally irradiated mice reconstituted with fetal liver. J Immunol 113:1353–1360
Siekevitz M, Huang SY, Gefter ML (1983) The genetic basis of antibody production: a single heavy chain variable region gene encodes all molecules bearing the dominant anti-arsonate idiotype in the strain A mouse. Eur J Immunol 13:123–131
Sigal N, Pickard AR, Metcalf ES, Gearhart PJ, Klinman NR (1977) Expression of phosphorylcholine-specific B cells during murine development. J Exp Med 146:933–948
Silverstein AM, Uhr JW, Kramer KL, Lukes RJ (1963) Fetal response to antigenic stimulus. II. Antibody production by the fetal lamb. J Exp Med 117:799–812
Smith RT, Eitzman DV, Catlin ME, Wirtz EO, Miller BE (1964) The development of the immune response. Characterization of the response of the human infant and adult to immunization with *Salmonella* vaccines. Pediatrics 33:163–183
Snodgrass HR, Kisielow P, Kiefer M, Steinmetz M, von Boehmer H (1985) Ontogeny of the T cell antigen receptor within the thymus. Nature 313:592–595
Stamatoyannopoulos G, Nakamoto B, Kurachi S, Papayannopoulou T (1983) Direct evidence for interaction between human erythroid progenitor cells and a hemoglobin switching activity present in fetal sheep serum. Proc Natl Acad Sci USA 80:5650–5654
Stein KE (1985) Network regulation of the immune response to bacterial polysaccharide antigens. Contemp Top Microbiol Immunol 119:57–78
Stohrer R, Kearney JF (1984) Ontogeny of B cell precursors responding to α-(1→3) dextran in BALB/c mice. J Immunol 133:2323–2326
Szewczuk MR, Siskind GW (1977) Ontogeny of B lymphocyte function. III. In vivo and in vitro studies on the ease of tolerance induction in B lymphocytes from fetal, neonatal and adult mice. J Exp Med 145:1590–1601
Teale JM (1985) B cell immune repertoire diversifies in a predictable temporal order in vitro. J Immunol 135:954–958
Teale JM, Kearney JF (1986) Clonotypic analysis of the fetal B cell repertoire. J Molec Cell Immunol 283–292
Tonegawa S (1983) Somatic generation of antibody diversity. Nature 302:575–581
Weaver D, Costantini F, Imanishi-Kari T, Baltimore D (1985) A transgenic immunoglobulin mu gene prevents rearrangements of endogenous genes. Cell 42:117–127
Weigert M, Riblet R (1978) The genetic control of antibody variable regions in the mouse. Springer Semin Immunopathol 1:133–169
Whitlock C, Denis K, Robertson D, Witte O (1985) In vitro analysis of murine B cell development. Annu Rev Immunol 3:213–235
Yancopoulos GD, Alt FW (1985) Developmentally controlled and tissue specific expression of unrearranged V_H genes segments. Cell 40:271–281
Yancopoulos GD, Alt FW (1986) Regulation of the assembly and expression of variable region genes. Annu Rev Immunol 4:339–368
Yancopoulos GD, Desiderio SV, Paskind M, Kearney JF, Baltimore D, Alt FW (1984) Preferential utilization of the most J_H-proximal V_H gene segments in pre-B-cell lines. Nature 311:727–733
Yancopoulos GD, Blackwell TK, Suh H, Hood L, Alt FW (1986) Introduced T cell receptor variable region gene segments recombine in pre-B cells: evidence that B and T cells use a common recombinase. Cell 44:251–259

Questioning the Role of the Embryonic Bursa in the Molecular Differentiation of B Lymphocytes

J.-C. WEILL[1], M. LEIBOWITCH[2], and C.-A. REYNAUD[1]

1 Introduction

The chicken bursa of Fabricius is a primary organ of differentiation for the B-cell lineage (GLICK et al. 1956). Its removal during embryonic development (up to 17 days of incubation) induces severe agammaglobulinemia and prevents the animal from mounting an immune response to any immunizing antigen (MUELLER et al. 1959; COOPER et al. 1969; WARNER et al. 1969). The stem cells which give rise to B lymphocytes colonize the bursa from the general circulation (MOORE and OWEN 1965, 1966; LE DOUARIN et al. 1975). They migrate at a specific stage of embryonic development to the bursal mesenchyme, where some may develop into granulocytes. When they reach the bursal epithelium, they induce the formation of bursal follicles and differentiate into B lymphocytes (HOUSSAINT et al. 1976).

Initially we were tempted to imagine the bursal differentiation stage as a typical instructive model in which a multipotent hematopoietic stem cell would be driven, depending on environmental influences, into the myeloid or the lymphoid pathway. By analogy with the mammalian immune system, we then expected the embryonic bursal follicle to be a berth where immunoglobulin (Ig) genes would rearrange in each newly generated B cell, giving random access at each division to a new heavy- and light-chain variable gene combination.

Our first results on the size of the pool of functional V genes in the chicken are in complete contradiction with this presupposition. The generation of diversity in this species is based essentially on somatic diversification mechanisms

[1] Laboratoire d'Immunodifférenciation moléculaire, Institut Jacques Monod, CNRS/Université Paris VII, 2, Place Jussieu, 75251 Paris Cedex 05, France
[2] Département de Dermatologie, Hôpital Tarnier, 89, Rue d'Assas, 75006 Paris, France

Current Topics in Microbiology and Immunology, Vol. 135

which operate on very few *V* sequences (Reynaud et al. 1985). The chicken humoral immune system does not have many functional *V* genes, but it has unusual ways of diversifying them. The bursa, as we show in this report, is not, then, the expected "rearranging machine". It is, however, the organ where lymphoid stem cells have to accumulate and mature before giving rise to a diversified population of B cells.

2 Rearrangement of Ig Genes in Lymphoid Tissue

We have isolated and characterized heavy (C_μ) and λ light chains ($C+V$) cDNA probes (Dahan et al. 1983; Reynaud et al. 1983). These probes have allowed us to study the various configurations of the Ig genes in chicken lymphoid tissues at different stages of development.

When looking at bursal DNA several weeks after hatching, we always found a single major rearrangement pattern of the light-chain genes compared to red blood cell DNA as germ-line reference (Fig. 1). After quantitation, 43% of

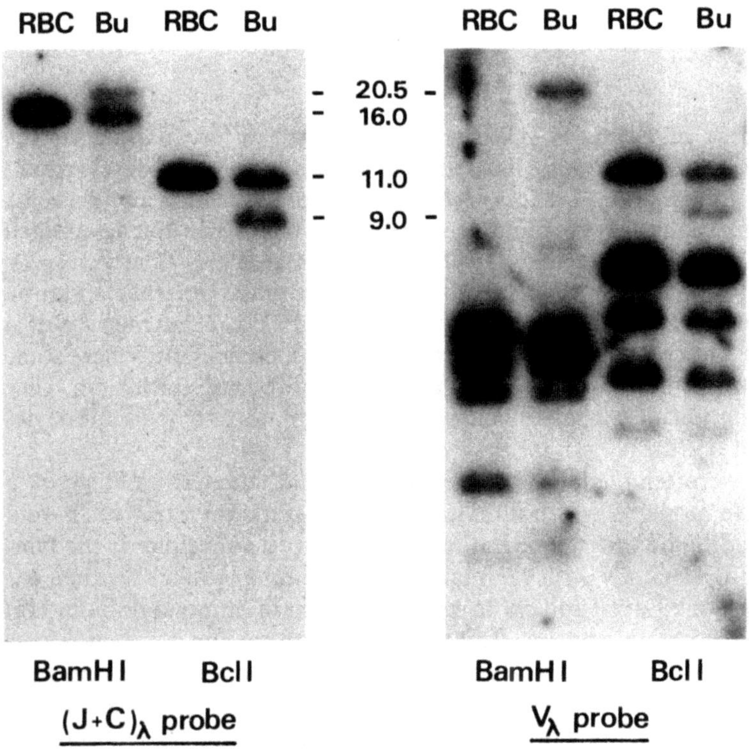

Fig. 1. Rearrangement of λ light-chain DNA sequences in the chicken bursa. Duplicate Southern blots of *Bcl*I- and *Bam*HI-digested DNA from red blood cells (*RBC*) and from bursal cells (*Bu*) taken from the same chicken were hybridized with a genomic $(J+C)_\lambda$ probe (left) or V_λ cDNA probe (*right*). The sizes (in kb) of germ-line ($J+C$) and rearranged ($V+J+C$)-hybridizing fragments are indicated

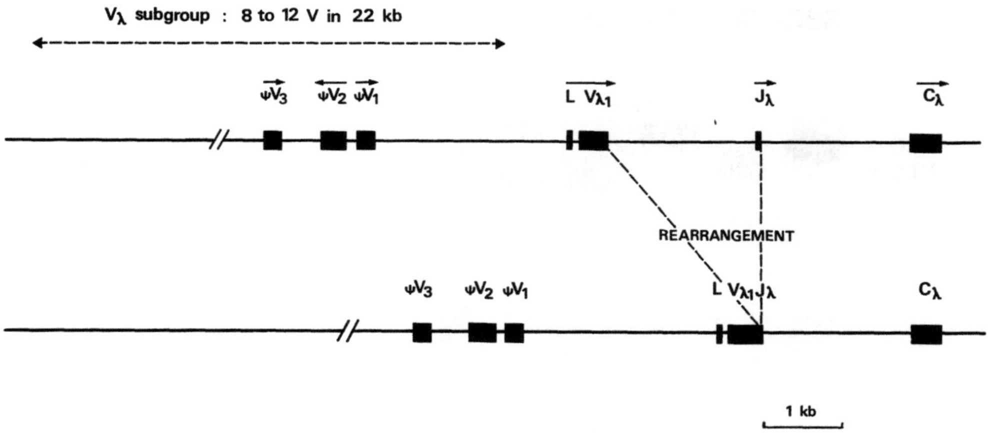

Fig. 2. Organization of the λ light-chain locus. The λ light chain locus is represented in its germ-line context and in its rearranged configuration in normal B cells. The *heavy boxes* represent the coding regions whose sequence was determined: the constant (*C*), joining (*J*), variable ($V_{\lambda 1}$), and leader (*L*) segments, and the three adjacent pseudogenes (ΨV_1, ΨV_2, ΨV_3). Horizontal arrows above these coding sequences indicate their transcriptional orientation. The whole V_λ subgroup (8–12 *V*-hybridizing elements) is clustered in a single 22-kb *BglI* fragment

the J_λ-hybridizing band was found in a rearranged configuration and 57% in the germ-line one. As 90% of the bursal cells were IgM-positive at the stage studied, these results implied that in almost all bursal cells, the first joining event involved the same light-chain $V_{\lambda 1}$ gene. When the chicken light-chain locus was studied, it showed a unique germ-line configuration, with a single constant region (C_λ), one junctional sequence (J_λ), and, almost adjacent to it (2 kb upstream), a functional V_λ gene ($V_{\lambda 1}$). Upstream from $V_{\lambda 1}$ lies the V_λ subgroup which contains 8–12 genes in a 22-kb cluster (Fig. 2). The sequence of the first three genes of this subgroup revealed that they are truncated genes which could not code for a functional variable domain. In the rearranged configuration in B-lymphoid cells, the sequence between V_λ and J_λ is deleted, thus allowing the joining of the two coding segments (Reynaud et al. 1985).

When probing bursal DNA with a C_μ probe, we also observed a major rearrangement event (Fig. 3). However, the μ and light-chain patterns differed in that the μ germ-line band disappeared almost completely, implying that, for this locus, both alleles may be rearranged, as is the case for the μ rearrangement in mouse splenic B cells (Nottenburg and Weissman 1981; Coleclough et al. 1981). When we used different restriction enzymes, a few minor bands very similar in size to the major rearranged fragment were always observed, creating a fuzziness on the autoradiogram next to the major rearranged band. This was also seen when bursal DNA was hybridized with a V_H probe. At present, we assume that the chicken uses one main V_H gene which is rearranged in most lymphoid bursal cells, the other minor events possibly being created by the use of other D_H or J_H fragments and by some nonproductive rearrangements.

These points remain speculative because of our lack of a genomic clone, the μ-coding sequence always suffering some large deletions during isolation.

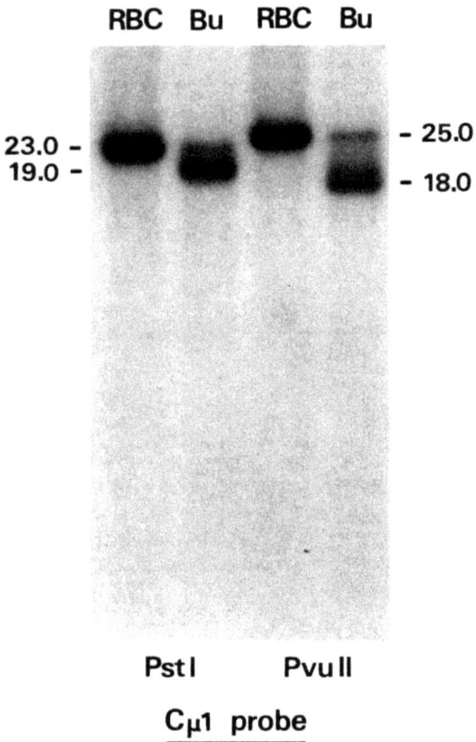

Fig. 3. Rearrangement of μ heavy-chain DNA sequences in the chicken bursa. Southern blots of *Pst*I- or *Pvu*II-digested DNA from red blood cells (*RBC*) or bursal cells (*Bu*) taken from the same chicken were hybridized with a $C_{\mu 1}$ cDNA probe. The sizes (in kb) of germ-line and rearranged $C_{\mu 1}$-hybridizing fragments are indicated

Such deletions have been described in the mouse switch region during the cloning of the μ locus, but never to such a large extent. They suggest the presence of several repeated sequences which are hotspots for recombination.

These major rearrangement events detected with the heavy and light chain genes in chicken lymphoid tissue can be observed in the bursa at the end of the embryonic period, during posthatching development, and also in the spleen of the adult animal.

3 Rearrangement of Ig Genes in Individual Bursal Follicles

Prebursal stem cells which can be detected at day 7 in the embryo colonize the bursal rudiment between days 8 and 14 of incubation. They appear as large basophilic cells that can be easily identified by a modified Giemsa coloration, which stains their cytoplasm dark blue (Houssaint et al. 1976) (Fig. 4). When they reach the bursal epithelium a budding of the epithelial layer occurs, creating

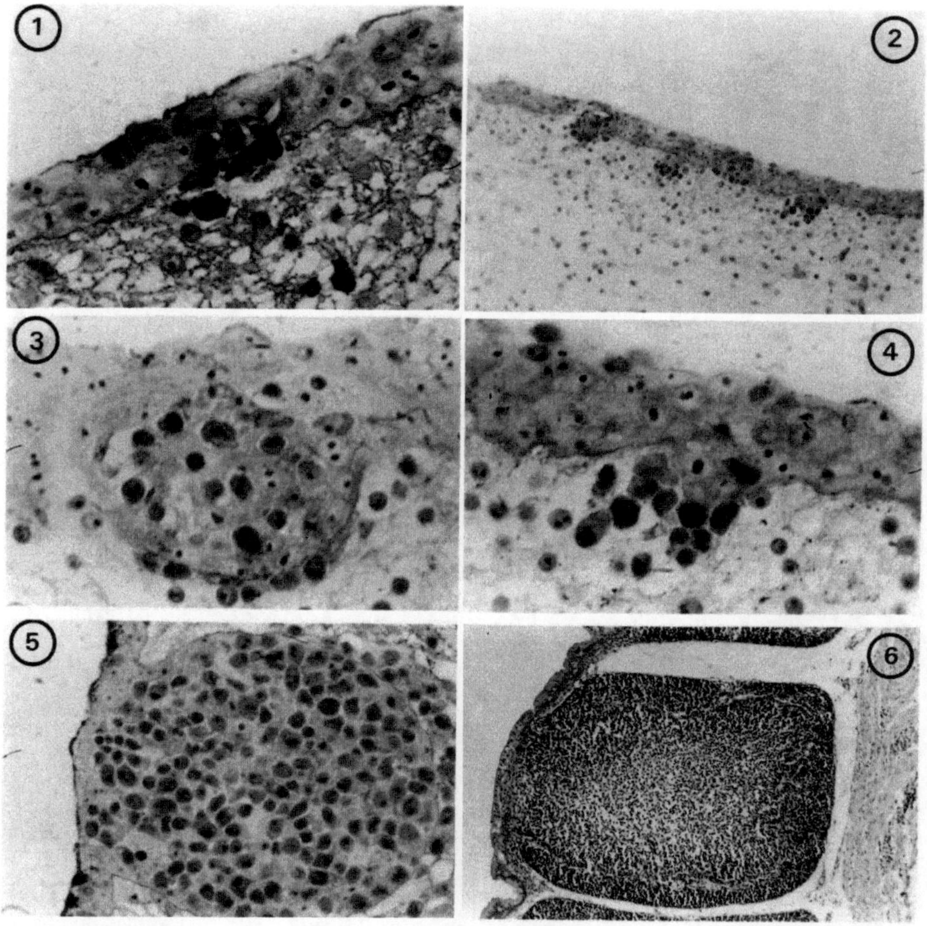

Fig. 4. Histological sections of the bursa of Fabricius at different developmental stages. Bursal tissue was embedded in resin and semi-thin sections (1 μm) (LESSANA-LEIBOWITCH et al. 1984) were submitted to the Pappenheim staining procedure (HOUSSAINT et al. 1976). Large basophilic stem cells can easily be detected at day 13–14 of embryonic development (*1–4*). At later stages the follicle appears full of lymphocytes (*5, 6*). *1* Day 13 of incubation, ×330; *2* day 14 of incubation, ×83; *3* day 14 of incubation, ×330; *4* day 14 of incubation, ×33; *5* day 18 of incubation, ×210; *6* 4 weeks after hatching, ×40

a follicular structure around these cells. Inside each follicle, these stem cells will divide and give rise to a population of cells bearing IgM at their surface, 90% of bursal cells being B lymphocytes at day 20 of embryonic development (EWERT et al. 1984). The bursa is composed of approximately 10^4 follicles (OLAH and GLICK 1978) which can be easily isolated (Fig. 5), giving access at different stages of development to pure segregated populations of B cells ($1–3 \times 10^5$ B cells per follicle at 3–8 weeks of posthatching development).

We have examined the rearrangement patterns of Ig genes in isolated bursal follicles and compared them with the corresponding pattern obtained from total bursal DNA. Such analysis should demonstrate whether each follicle shows

Fig. 5. Macroscopic view of a bursa at 6 weeks after hatching with its typical structure in plicae. Individual follicles have been released into the medium

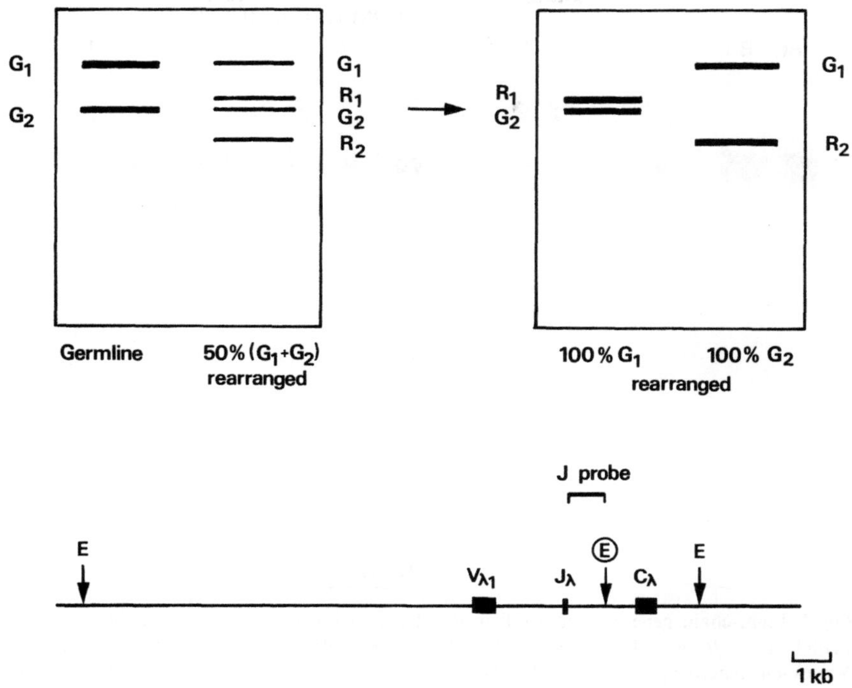

Fig. 6. Schematic representation of the different configurations of the light-chain locus in heterozygous birds. *Bottom:* The C_λ genomic locus with C_λ, J_λ, and $V_{\lambda 1}$ exons. *Eco*RI (*E*) restriction sites are indicated. The polymorphic site is circled. *Top:* Hybridization patterns of *Eco*RI digests probed with a *J* region fragment. *Left: Eco*RI digests of germ-line (red blood cell) and rearranged (bursal) DNA. The germ-line configuration shows the two alleles G_1 and G_2. The rearranged configuration shows G_1 and G_2 together with two rearranged fragments R_1 and R_2. If both alleles are rearranged in a random fashion, these four fragments should have the same intensity (half of the germ-line intensity G_1 and G_2). *Right:* The same experiment on two different monoclonal populations of B cells. In this case all cells have rearranged either G_1 or G_2 and show an R_1G_2 or a G_1R_2 pattern

a monoclonal or a diversified pattern of rearrangement and allows us to evaluate the average number of rearrangement events taking place in a follicle.

Having reported that the same rearrangement event involving a unique functional V_λ gene occurred in most chicken B cells, we had to bypass this obvious obstacle and therefore took advantage in our experimental design of the presence of a polymorphic *Eco*RI site located in the light-chain locus. In Fig. 6 we show the general outline of this approach. Analysis showed that a large number (27/53) of individual follicles presented one rearrangement event. Very few follicles presented a typical mixed rearranged configuration as observed with the corresponding total bursa, and most of the remaining follicles showed a partially mixed pattern more easily explained by an unequal representation of the two forms of events (Fig. 7).

When the μ heavy-chain locus was analyzed in the same way, a great number of follicles once again showed a simplified pattern as compared to the total bursa (Fig. 8). Our data on the heavy- and light-chain gene rearrangements

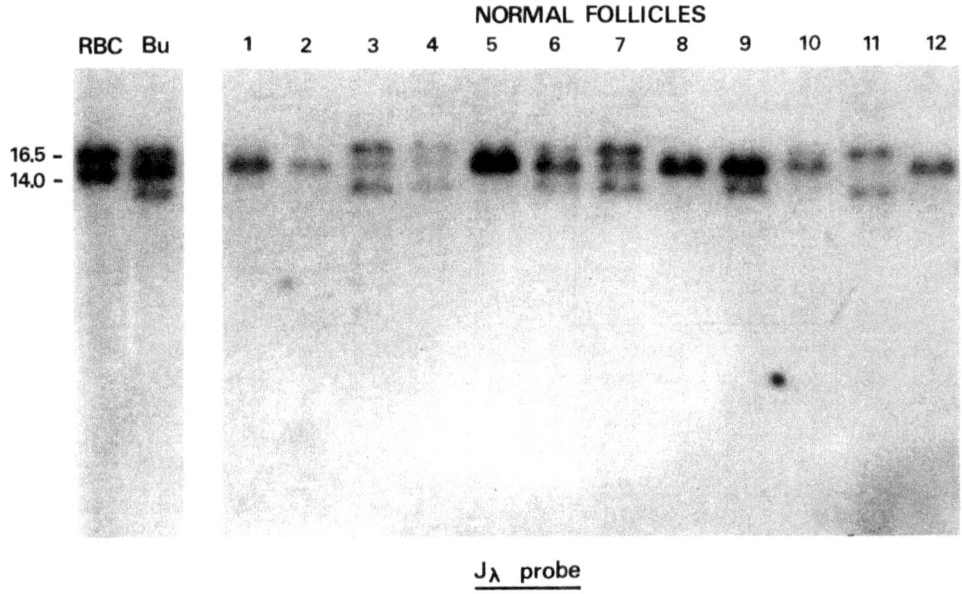

Fig. 7. Light-chain gene configuration in isolated follicles from normal birds heterozygous for the *Eco*RI site. *Left: Eco*RI digest of red blood cell DNA (*RBC*) and bursal DNA (*Bu*) from a normal 6-week-old heterozygous animal. The two germ-line bands are shown at 16.5 kb and 14 kb. *Right, 1–12: Eco*RI digest of DNA from single follicles isolated from a normal 6-week-old heterozygous bird. All the blots were hybridized with the *J* probe (see Fig. 6)

in isolated follicles are thus concordant. They allow us, on the basis of the light-chain result and using a statistical calculation, to estimate the average number of rearrangements as two events per follicle (Weill et al. 1986).

These results favor a model where each follicle is colonized by a small number of prebursal stem cells. These cells are committed to a particular Ig gene rearrangement at the very beginning of the development of the embryonic bursa and do not continually rearrange their Ig genes during the maturation of this organ. Such a model of development for the chicken immune system implies that the rearrangement is a unique event in the life of the chicken. It takes place during a short period of embryonic development, but we do not yet know whether this period occurs at the very beginning of the colonization process, either in the mesenchyme or at the onset of follicular formation, or possibly even before the basophilic stem cells enter the bursa. This last proposition would suggest that different progenitors for the separate hematopoietic lineages may already be present at day 7 of incubation in the general circulation.

Our results extend our previous densitometric analysis, which implied that in chicken B cells, the most proximal V_λ gene appeared always to be productively rearranged on one allele, the other allele remaining in the germ-line configuration (Reynaud et al. 1985). In fact, among all the follicles analyzed so far (100), only one displayed a rearrangement event on both chromosomes (see Fig. 7, follicle 9, showing the loss of the upper 16.5-kb germ-line fragment).

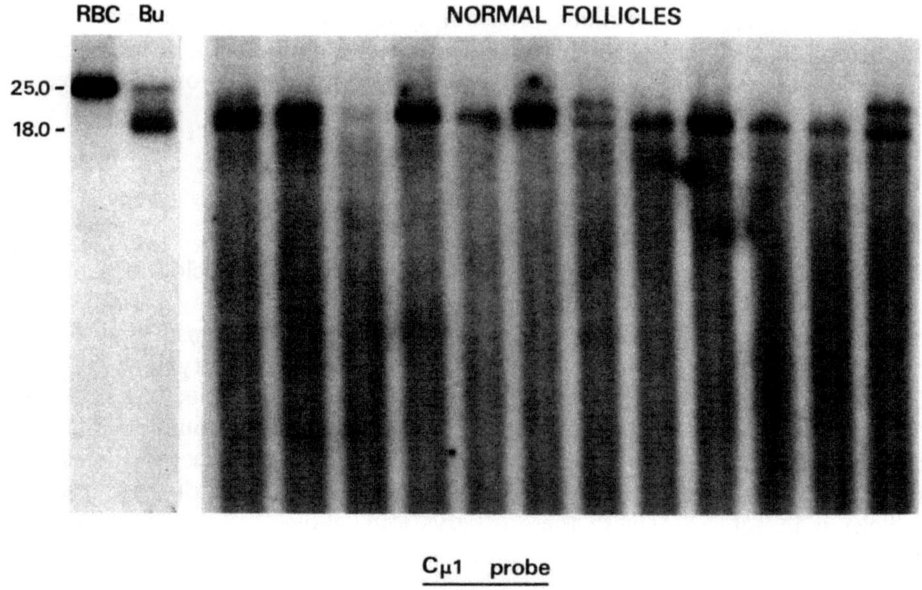

Fig. 8. Heavy-chain gene configuration in isolated follicles from normal birds. *Left: Pvu*II digest of normal red blood cell DNA (*RBC*) and bursal DNA (*Bu*). The 25-kb germ-line band and the major rearranged fragment at 18 kb are shown. *Right: Pvu*II digest of DNA from single follicles isolated from a normal 6-week-old bird. All the blots were hybridized with a 500-bp-$C_{\mu 1}$ subfragment (WEILL et al. 1986)

It remains to be explained by what mechanism the V_λ–J_λ joining could be achieved with such precision. There may be, as in mammals, two chances in three of creating an aborted joint, corresponding to the probability of joining two segments of DNA while maintaining the reading frame. The same probability would apply to the heavy-chain locus, and in such a case one stem cell in nine would succeed in rearranging its heavy- and light-chain loci productively. As heavy- and light-chain genes are expressed and probably rearranged concomitantly in birds (GROSSI et al. 1977), we would like to propose that this event will generally be attempted only once on one allele in bursal lymphoid stem cells. In two-thirds of the cases at each locus, the joining will be out of frame and the cells will be discarded. In such cases, it will take an average of 18 basophilic stem cells to obtain two productively rearranged lymphoid precursors in each bursal follicle. Accordingly, when looking at follicular sections at day 13–14 of embryonic development, a rather large number of basophilic cells can be detected (Fig. 4). However, these cells could also be functionally rearranged stem cells which would first divide in each follicle before differentiating into B lymphocytes. It remains very difficult at the moment to correlate the basophilic phenotype with a specific differentiation stage.

In contrast to the strictly time-dependent nature of this event in birds, the rearrangement mechanism in mammalian species has evolved into a steady process, probably occurring continually in the bone marrow. It may therefore involve a more precisely regulated program, following a hierarchical order from the heavy- to the light-chain locus and using a trial-and-error mechanism to proceed from one allele to the other in order to produce a functional Ig molecule (ALT et al. 1986).

4 The Committed Stem Cells for the B Lineage

It has been demonstrated that the B-lymphoid system of cyclophosphamide-treated, B-cell-depleted, 4-day-old chicks can be reconstituted by normal bursal cells of the same age (TOIVANEN and TOIVANEN 1973). Such reconstituted follicles have been shown to be of monoclonal origin at the phenotypic level, each follicle being populated by a single lymphoid stem cell (PINK et al. 1985a). Moreover, the reconstitution of these depleted bursal follicles could be abolished by prior incubation of the donor cells with an antiserum directed against the surface-IgM molecule (PINK et al. 1985b). The DNA analysis of these restored follicles is in agreement with these observations, showing that the single stem cell which can repopulate (up to 3×10^5 cells) a depleted follicle 4 days after hatching, is already committed to a particular Ig gene rearrangement (WEILL et al. 1986).

It has also been shown that long-lasting Ig suppression could be obtained by injection of 13-day-old chick embryos with an anti-IgM allotype serum (RATCLIFFE and IVANYI 1981). This suggests that the surface-IgM stem cell which is the target of the anti-μ suppression would appear in the embryonic bursa during the period of colonization ($1–4 \times 10^4$ stem cells at day 8–14 of incubation). Several weeks after hatching, these bursal stem cells reach the postbursal stage where they can restore a B-cell-depleted animal without homing to the bursa (TOIVANEN and TOIVANEN 1973). This pool of extrabursal committed stem cells is then responsible for the renewal of the B-cell population for the rest of the animal's life. At this stage, B cells cannot be newly generated from an early multipotent stem cell, for the bursa has involuted and can no longer provide the environment necessary for this early step to occur.

We have previously suggested that the generation of chicken light-chain diversity may rely chiefly on some precise somatic diversification mechanisms in the developing bursa (REYNAUD et al. 1985). An identical picture seems to emerge for the heavy-chain locus. This diversification program operates at the IgM level and appears indispensable for the generation of a large preimmune repertoire in this species. It appears, then, that the rearrangement of Ig genes must be able to trigger the diversification program, these two connected mechanisms most probably involving two different sets of regulatory proteins and enzymes.

In fact, when rearranged $V_{\lambda 1}$ sequences are examined at different stages of bursal development, they present very precise and localized modifications, whereas the unrearranged V_λ sequence which is 2 kb 5′ of the J_λ region on the other allele remains unchanged (mansucript in preparation).

Several points concerning the molecular behavior of the bursal stem cells during the diversification program remain to be elucidated. Do these cells keep a germ-line V_H and V_L segment at their surface, the diversification mechanisms operating only on their descendants? Or do they accumulate many different somatic modifications during the bursal stage, therefore presenting a rather large immune repertoire when they seed away to the periphery? The first proposition would imply that the diversification program can be turned on in the progeny of these stem cells outside the bursa during the adult life of the animal. In the second hypothesis, the bursa would remain the only place where the diversification program can be activated. In both cases, expansion of B-cell clones after stimulation by antigen could induce a whole series of somatic mutations as described in the mouse.

5 Avian V_s Mammalian B-Cell Development

It has been shown that there is a progressive emergence in the mouse fetal liver of a stem cell population which can give rise to B lymphocytes. This differentiation does not seem to be closely related to the appearance of cells detected by the in vivo spleen focus assay (CFU-S). These committed B-cell progenitors are present in fetal liver and in adult bone marrow and are able to generate B cells in partially immunodeficient CBA/N mice 1–2 weeks after their injection (Paige et al. 1981). They do not carry Ig at their surface and may be distinguished from multipotent stem cells present at an early stage of embryonic life. Indeed, in a similar reconstitution experiment, 9 weeks are required for the formation of B cells after the injection of 9-day-old yolk sac cells (Paige et al. 1979). Precursors present in long-term mouse bone marrow cultures can also restore the presence of clonable B cells (CFU-B) in CBA/N mice in 2–6 weeks, whereas no CFU-S appear in the spleen of these animals (Kurland et al. 1984). A major prediction from these different reports is that committed progenitors for the B-cell system may be formed early in development in mammals and constitute the pool of precursors which will generate the B-lymphocyte population during the life of the animal. This proposition remains controversial, but experiments involving introduction of selectable markers into multipotent stem cells may eventually allow the issue to be settled.

It remains striking, however, that the ontogenetic pathway which we wish to propose for the chicken B-cell development involves the formation during embryonic development of a population of committed progenitors for the B-cell lineage. If a similar general mechanism has been conserved for the mammalian B-cell development, one main difference would emerge between these two models. In the mouse, the committed stem cells have probably not rearranged their V genes and will do so at each division in the fetal liver and later in the bone marrow. In the chicken, these cells are surface-IgM-positive; they will be present first in the bursa, from the embryonic period until 4–8 weeks of life, and then in peripheral lymphoid tissues. Such a general differentiation pathway for the B-cell lineage would, in avian species, strictly require a bursal stage at a precise time of development allowing for the segregation of B-cell progenitors. The mammalian pathway could follow the same general rule, but it would also provide the possibility of falling back at any time upon a multipotent stem cell. Such a feedback mechanism could then replenish a pool of intermediate progenitors in specific situations of demand.

6 Avian Leukosis Virus Infection: A Bursa-Dependent Lymphoid Tumor

The bursa of Fabricius is the site of primary tumor development in chickens susceptible to avian leukosis virus (ALV)-induced B-cell lymphoma. Early removal of the bursa prevents formation of this tumor, and it was suggested many years ago that this could be related to the key role played by the bursa in the differentiation of B lymphocytes (Cooper et al. 1968). At the molecular level, ALV infection results in the insertion of viral sequences (LTR) adjacent to the cellular *myc* gene. This insertion deregulates the transcription of the c-*myc* gene, this event possibly being responsible for the proliferation of the target cell (Hayward et al. 1981). Following ALV infection, the bursa contains hyperplastic preneoplastic follicles which, when examined at the histological level, contain mostly large undifferentiated pyrinophilic lymphoblasts (Cooper et al. 1968; Baba and Humphries 1985).

During a classical restoration experiment (see Sect. 4) in which bursal cells restore a cyclophosphamide-depleted bursa, these large pyrinophilic lymphoblasts first appear along the basement membrane of the depleted follicle and then give rise to more mature lymphocytes, as shown by Neiman et al. (1985). These authors have elegantly demonstrated that bursal cells, prior to their injection into a cyclophosphamide-depleted host, could be infected with a myelocytomatosis virus (HB21) containing a v-*myc* oncogene. After incubation and integration of the v-*myc* gene, the bursal cells were still able to restore the depleted bursa, but a large proportion (up to 50% in some cases) of the restored follicles were populated with large pyrinophilic undifferentiated lymphoblasts, these being very similar to cells present in the preneoplastic hyperplastic follicles observed after ALV infection. Putting these observations together, it is tempting to propose that the target of the ALV infection and of the HB21 v-*myc*-containing virus in Neiman et al.'s experiment may be the bursal lymphoid stem cell. This hypothesis also implies, mainly on a histological basis, that the bursal lymphoid stem cell, once infected by the virus, is blocked in its differentiation pathway and will populate the follicle it has colonized with identically arrested cells. The existence of the bursa may then be necessary for ALV infection because it is the organ where bursal stem cells accumulate in a compartmentalized anatomical location and where they can be reached by the infectious agent. These cells are present in peripheral organs in the adult animal, but may then be more disseminated or may no longer be susceptible to viral penetration. In this sense these hyperplastic follicles could be a source of bursal stem cells allowing, at different stages, molecular analysis of their Ig receptors.

Acknowledgements. We would like to thank Miss Lamy for expert technical assistance with the histological preparations. This work was supported by grants from the Centre National de la Recherche Scientifique, the Association pour la Recherche contre le Cancer, and the Institut National de la Santé et de la Recherche Médicale.

References

Alt FW, Blackwell TK, de Pinho RA, Reth MG, Yancopoulos GD (1986) Regulation of genome rearrangement events during lymphocyte differentiation. Immunol Rev 89:5–30

Baba TW, Humphries EH (1985) Formation of a transformed follicle is necessary but not sufficient for development of an avian leukosis virus-induced lymphoma. Proc Natl Acad Sci USA 82:213–216

Coleclough C, Perry RP, Karjalainen K, Weigert M (1981) Aberrant rearrangements contribute significantly to the allelic exclusion of immunoglobulin gene expression. Nature 290:372

Cooper MD, Payne LN, Dent PB, Burmester BR, Good A (1968) Pathogenesis of avian lymphoid leukosis. I. Histogenesis. JNCI 41:373–383

Cooper MD, Cain WA, van Alten PJ, Good RA (1969) Development and function of the immunoglobulin producing system. Int Arch Allergy 35:242–252

Dahan A, Reynaud CA, Weill JC (1983) Nucleotide sequence of the constant region of a chicken μ heavy chain immunoglobulin. Nucleic Acids Res 11:5381–5389

Ewert DL, Munchus MS, Chen CLH, Cooper MD (1984) Analysis of structural properties and cellular distribution of avian Ia antigen by using monoclonal antibody to monomorphic determinants. J Immunol 132:2524–2530

Glick B, Chang TS, Jaap RG (1956) The bursa of Fabricius and antibody production. Poultry Sci 35:224

Grossi CE, Lydyard PM, Cooper MD (1977) Ontogeny of B cells in the chicken. II. Changing patterns of cytoplasmic IgM expression and of modulation requirements for surface IgM by anti-μ antibodies. J Immunol 119:749–756

Hayward WS, Neel BG, Astrin SM (1981) Activation of a cellular *onc* gene by promoter insertion in ALV-induced lymphoid leukosis. Nature 290:475–480

Houssaint E, Belo M, Le Douarin NM (1976) Investigations on cell lineage and tissue interactions in the developing bursa of Fabricius through interspecific chimeras. Dev Biol 53:250–264

Kurland JI, Ziegler SF, Witte ON (1984) Long-term cultured B-lymphocytes and their precursors reconstitute the B-lymphocyte lineage in vivo. Proc Natl Acad Sci USA 81:7554–7558

Le Douarin NM, Houssaint E, Jotereau FV, Belo M (1975) Origin of hematopoietic stem cells in embryonic bursa of Fabricius and bone marrow studied through interspecific chimeras. Proc Natl Acad Sci USA 72:2701–2705

Lessana-Leibowitch M, Prado A, Palangie A, Lamy F, Flandrin G (1984) The diagnosis of cutaneous T-cell lymphoma by morphometric evaluation of the cellular infiltrate using semithin sections. Br J Dermatol 110:511–521

Moore MAS, Owen JJT (1965) Chromosome marker studies on the development of the haematopoietic system in the chick embryo. Nature 208:956–957

Moore MAS, Owen JJT (1966) Experimental studies on the development of the bursa of Fabricius. Dev Biol 14:40–51

Mueller AP, Wolfe HR, Meyer RK (1959) Precipitin production in chickens. XXI. Antibody production in bursectomized chickens and in chickens injected with 19-nortestosterone on the fifth day of incubation. J Immunol 83:507–510

Neiman P, Wolf C, Enrietto PJ, Cooper GM (1985) A retroviral *myc* gene induces preneoplastic transformation of lymphocytes in a bursal transplantation assay. Proc Natl Acad Sci USA 82:222–226

Nottenburg C, Weissman IL (1981) Cμ gene rearrangement of mouse immunoglobulin genes in normal B cells occurs on both the expressed and non expressed chromosomes. Proc Natl Acad Sci USA 78:484

Olah I, Glick B (1978) The number and size of the follicular epithelium (FE) and follicles in the bursa of Fabricius. Poultry Sci 57:1445–1450

Paige CJ, Kincade PW, Moore MAS, Lee G (1979) The fate of fetal and adult B-cell progenitors grafted into immunodeficient CBA/N mice. J Exp Med 150:548–563

Paige CJ, Kincade PW, Shinefeld LA, Sato VL (1981) Precursors of murine B lymphocytes. Physical and functional characterization, and distinction from myeloid stem cells. J Exp Med 153:154–165

Pink JR, Vainio O, Rijnbeek AM (1985a) Clones of B lymphocytes in individual follicles of the bursa of Fabricius. Eur J Immunol 15:83–87

Pink JR, Ratcliffe MJH, Vainio O (1985b) Immunoglobulin-bearing stem cells for clones of B (bursa-derived) lymphocytes. Eur J Immunol 15:617–620

Ratcliffe MJH, Ivanyi J (1981) Allotype suppression in the chicken. IV. Deletion of B cells and lack of suppressor cells during chronic suppression. Eur J Immunol 11:306–310

Reynaud CA, Dahan A, Weill JC (1983) Complete sequence of a chicken light chain immunoglobulin derived from the nucleotide sequence of its mRNA. Proc Natl Acad Sci USA 80:4099–4103

Reynaud CA, Anquez V, Dahan A, Weill JC (1985) A single rearrangement event generates most of the chicken immunoglobulin light chain diversity. Cell 40:289–291

Toivanen P, Toivanen A (1973) Bursal and postbursal stem cells in the chicken. Functional characteristics. Eur J Immunol 3:585–595

Warner NL, Uhr JW, Thorbecke GJ, Ovary Z (1969) Immunoglobulins, antibodies and the bursa of Fabricius: induction of agammaglobulinemia and the loss of all antibody-forming capacity by hormonal bursectomy. J Immunol 103:1317

Weill JC, Reynaud CA, Lassila O, Pink JRL (1986) Rearrangement of chicken immunoglobulin genes is not an ongoing process in the embryonic bursa of Fabricius. Proc Natl Acad Sci USA 83:3336–3340

Immunoglobulin Gene Rearrangements in Pre-B Cells

P.D. BURROWS and H. KUBAGAWA

1 Introduction: The Cells

Although the existence of antibodies and their induction by antigen had been shown by von Behring and Kitasato in the 1890s, the cellular origin of antibodies remained controversial right up to the late 1940s, when it was determined that plasma cells contained large amounts of immunoglobulin (Ig). It was subsequently shown that the progenitor of the plasma cell was a small lymphocyte and that the lymphocyte precursors of plasma cells represented a separate differentiation pathway of cells derived from the avian bursa or mammalian bone marrow (for review see COOPER and LAWTON 1974). These B lymphocytes were found to bear membrane-bound Ig as a receptor for antigen (RAFF et al. 1970; PERNIS et al. 1970). The identification of the precursors of B lymphocytes (see review KINCADE 1981; COOPER 1981) proved to be difficult, since it relied on indirect assays, e.g., the ability of fractionated cell populations to differentiate into B lymphocytes after transfer into irradiated hosts, or the in vitro generation of B lymphocytes in organ cultures of fetal liver (e.g. OSMOND and NOSSAL 1974; OWEN et al. 1974; RYSER and VASSALLI 1974; MELCHERS et al. 1976). These types of experiments were important in defining the existence and tissue distribution of precursors of B lymphocytes, but were handicapped by the inability to identify the precursor cells directly as well as functionally. The identification of these precursors became possible when it was discovered that cells con-

Cellular Immunobiology Unit, Department of Microbiology, and
The Comprehensive Cancer Center, University of Alabama at Birmingham, Birmingham, Alabama 35294, USA

Current Topics in Microbiology and Immunology, Vol. 135

taining intracellular IgM could be found in sites of B-cell generation, even prior to the appearance of B lymphocytes (RAFF et al. 1976). The finding that these cells synthesize only μ heavy chain in the absence of light-chain synthesis (BURROWS et al. 1979; SIDEN et al. 1979; LEVITT and COOPER 1980) made it clear that the pre-B cell was a unique differentiation stage in the B-cell lineage. Subsequent molecular analyses of pre-B-cell-derived hybridomas (MAKI et al. 1980; PERRY et al. 1981) and transformed cell lines (ALT et al. 1981) and of normal pre-B cells (COFFMAN and WEISSMAN 1984) revealed that no light chains were produced because the light-chain gene loci were in embryonic configuration, i.e., they had not yet undergone the somatic DNA rearrangements necessary for the expression of a light-chain gene. In this article we will discuss the regulation of the Ig gene rearrangements that may occur in pre-B cells and the consequences of defects in these rearrangement processes.

2 Regulation of Ig Gene Rearrangements in Pre-B Cells

A summary (Fig. 1) of the types of Ig gene rearrangements that occur in the heavy- and light-chain gene loci (reviewed in TONEGAWA 1983; HONJO 1983) reveals several steps where regulation of the process could occur, as well as points where a single genetic defect could abort the entire B-cell differentiation pathway and perhaps also the other arm of specific immunity, the T-cell lineage.

2.1 Heavy-Chain Gene Rearrangement

The variable region of the Ig heavy chain (V_H) is encoded by three gene segments – *V* (variable), *D* (diversity), and *J* (joining) – which exist together as a functional exon only in cells of B lineage. In other somatic cells and in germ cells these families of gene segments are widely separated on the DNA. Their juxtaposition requires rearrangement and deletion of the DNA, forming the *VDJ* exon which is transcribed with the μ constant region exons (C_μ) to generate μ heavy-chain mRNA. Studies of transformed cells and normal splenic B lymphocytes (SAKANO et al. 1981; SUGIYAMA et al. 1983; ALT et al. 1984; YAOITA et al. 1983) indicate that this rearrangement process is orderly and well regulated and suggest the following scenario for heavy-chain gene rearrangements: In the mouse, $D \rightarrow J$ rearrangements are the first to occur and do so on both homologues of chromosome 12, resulting in the genotype DJ/DJ. The cell then undergoes a $V \rightarrow DJ$ rearrangement on one allele. A successful rearrangement (VDJ^+/DJ) results in production of a complete μ heavy-chain protein, defining the cell as pre-B, and inhibits further rearrangement of the heavy-chain locus. Should a nonproductive, abortive rearrangement (VDJ^-/DJ) occur, the process can proceed on the other allele. If successful (VDJ^-/VDJ^+), the ensuing production of μ-chain will then again define the cell as a pre-B cell, indistinguishable in terms of Ig phenotype from the VDJ^+/DJ pre-B cells. Transformed cells with double abortive (VDJ^-/VDJ^-) rearrangements have been described (e.g., RETH et al. 1985; HAGIYA et al. 1986), and such cells are presumably generated during normal pre-B cell differentiation. However, the absence of μ-chain production would preclude their ready identification as pre-B cells. (The issue of whether such cells could undergo light-chain gene rearrangements, becoming μ-chain-negative, light-chain-positive "pre-B" cells, is discussed below.) Depending on the frequency of abortive rearrangements, there could exist a sizeable graveyard for these inept pre-B cells in sites of B-cell generation. Determination of the exact frequency of abortive rearrangements at the heavy-chain locus has been difficult. Early estimates indicated a very high frequency and were consistent with stochastic models for allelic exclusion, i.e., the frequency of productive rearrangements would be so low that the probability of a cell undergoing two productive heavy-chain rearrangements would be minuscule and compatible with the central dogma of immunology – one B cell, one Ig specificity. Such profligate behavior on the part of pre-B cells in the name of allelic exclusion is now thought not to occur. Earlier estimates of the frequency of abortive rearrangements were inflated, since (1) the probes used to analyze the rearranged heavy-chain genes could not distinguish between *DJ*, an intermediate in the rearrangement process, and VDJ^-, a true abortive rearrangement; and (2) in many murine plasmacytomas, a cell type used extensively in such analyses, the nonproductive allele has undergone a chromosomal translocation bringing the c-*myc* proto-oncogene into the Ig heavy-chain gene locus (reviewed in KLEIN 1983; LEDER et al. 1983; CROCE and NOWELL 1985). This type of translocation, which may involve the switch region, would also have been scored as an abortive rearrangement.

A common feature of the regulation of rearrangements at both heavy- and light-chain (see below) gene loci is that the cell can apparently "sense" success. Two likely candidates for the product of a productive heavy-chain rearrangement that can be sensed by the cell are μ RNA and μ protein. Several lines of evidence suggest that μ RNA cannot be the signaling molecule. Considerable amounts of μ RNA transcripts can be found in cells prior to completion of a productive rearrangement (KEMP et al. 1980; ALT et al. 1982b; NELSON et al. 1983; RETH and ALT 1984). These are generated from *DJ* rearrangements or are sterile "I" transcripts containing C_μ that initiate from pseudopromoters in the J–C_μ intron. At least one case has been documented where a VDJ^- rearrangement in an Abelson murine leukemia virus (A-MuLV)-transformed pre-B cell line was abortive because it encoded an inframe translational stop codon (ALT et al. 1982b; WABL et al. 1985). The other allele was productively rearranged. The μ message produced from the abortive allele would presumably be indistinguishable from a productive μ message. Only the absence of a complete μ heavy-chain protein would have marked the rearrangement as abortive. An assumption in this case is that the rearrangement history of the cell line proceeded $VDJ^-/DJ \rightarrow VDJ^-/VDJ^+$ and not $VDJ^+/DJ \rightarrow VDJ^+/VDJ^-$. This assumption is unfortunately not testable, but is consistent with data obtained from other A-MuLV-transformed cell lines. There is other negative evidence implicating μ heavy-chain protein as the indicator molecule for successful rearrangement. The RNA transcribed from *DJ* rearrangements is sometimes translated into a truncated μ protein (RETH and ALT 1984). However, production of high levels of this protein seems to inhibit $V \rightarrow DJ$ rearrangement in those A-MuLV lines capable of this process (RETH et al. 1985). The sterile μ transcripts initiating in the J–C_μ intron are apparently spliced in such a way as to create a new, nontranslatable exon, a "nontron," at the 5′ end of the μ mRNA (LENNON and PERRY 1985). Thus, the transcription of non- or incompletely rearranged heavy-chain genes is a common occurrence and may even be a prerequisite for rearrangement. However, only a fraction of these transcripts are ever translated, and the truncated μ protein is rarely expressed at high levels since this apparently inhibits rearrangement. An elegant method for testing the hypothesis that production of a heavy- or light-chain protein prevents further rearrangement of the respective loci involves creation of transgenic mice containing already rearranged heavy- and/or light-chain genes in their germ-line DNA. One would predict that mice with a rearranged heavy-chain gene would exclusively express the microinjected heavy-chain *V* region in association with the normal repertoire of κ and λ light-chain *V* region genes. Mice with both rearranged μ heavy- and κ light-chain genes, on the other hand, should be agammaglobulinemic save for the Ig produced by the microinjected genes, since rearrangement of all endogenous Ig genes would have been inhibited. Such experiments have been performed and the results have not been as straightforward as would have been expected from the simple predictions (RITCHIE et al. 1984; RUSCONI and KOEHLER 1985; STORB et al. 1985; WEAVER et al. 1985); rather, they are suggestive of an inhibitory effect of the microinjected Ig gene on endogenous Ig gene rearrangement.

2.2 Light-Chain Gene Rearrangements

In addition to the negative signal inhibiting further heavy-chain gene rearrangement, there is correlative evidence that a productive *VDJ* rearrangement can provide a positive signal, allowing the pre-B cell to proceed to the next stage of differentiation where light-chain gene rearrangement can occur.

In studies of A-MuLV-transformed cells lines that could undergo both heavy- ($V \rightarrow DJ$) and light-chain gene rearrangement, RETH et al. (1985) found that clones with double abortive rearrangements of the heavy-chain locus maintained their κ light-chain genes in germ-line configuration. With one interesting exception, all cells that had rearranged and expressed a κ light-chain gene had a productively rearranged heavy-chain gene. While synthesis of a complete μ chain may provide a positive signal leading to light-chain gene rearrangement, μ chain production in and of itself is not sufficient: pre-B-cell-derived hybridomas, Epstein–Barr virus (EBV)-transformed pre-B cells, and human and most murine pre-B cell leukemias produce significant amounts of μ heavy chain, but the light-chain loci remain germ-line. Some additional signal, either specific to differentiation stage and intrinsic to the pre-B cell or perhaps provided by the inductive microenvironment of the bone marrow, seems to be required for the induction of light-chain gene rearrangement.

This proposed dual role, positive and negative, for μ chain protein has several implications for B-cell differentiation, the most straightforward being that pre-B cells producing light chain without a heavy chain should not exist. In an extensive screening of normal bone marrow cells and fetal liver-derived hybridomas we have never found light-chain-only pre-B cells (KUBAGAWA et al. 1982; BURROWS and KEARNEY, unpublished). However, one study reported the isolation of light-chain-only hybridomas from fusions with murine fetal liver, and some rare A-MuLV transformants apparently synthesize only light-chain (SIDEN et al. 1979). Recently, we have been using EBV transformation of bone marrow cells to study human B-lineage differentiation. When analyzed by immunofluorescence, the transformed cells display a spectrum of Ig phenotypes, including light-chain-positive, heavy-chain-negative (KUBAGAWA et al. 1986). The nature of these light-chain-only cells is currently under investigation. They could be an artifact of EBV transformation. If so, the artifact must be tissue-specific, since in a preliminary screening these cells were seen only in bone marrow cultures, and not in cultures from blood mononuclear cells. It may be that these cells are faithful analogs of cells that exist but are not readily detectable in normal lymphopoietic tissue, perhaps being an abortive differentiation pathway. The critical question is whether the cells have rearranged their light-chain genes before or after heavy-chain gene rearrangement. If the cells have double abortive heavy-chain alleles, this would indicate that production of a complete μ heavy-chain protein is not an absolute prerequisite for light-chain gene rearrangement. Such cells in normal bone marrow would be a "dead end" differentiation stage since they presumably could never make a functional receptor for antigen. On the other hand, if the heavy-chain genes in these light-chain-only cells are in germ-line configuration, this would indicate that the rearrangement hierarchy, heavy before light, is not inviolate.

A cell that has successfully completed this first series of Ig gene rearrangements at the heavy-chain locus will be identifiable by immunologists as a pre-B cell, but is immunologically incompetent since the μ heavy chain produced remains intracellular. The expression of a surface receptor for antigen requires the rearrangement and expression of a light-chain gene. The DNA rearrangements involved in light-chain expression are simpler than those for heavy chains, requiring only a single $V_L \rightarrow J_L$ recombination event (reviewed in TONEGAWA 1983). However, the existence of two light-chain isotypes, κ and λ, adds another level of complexity to the regulation of the rearrangement process. To maintain monospecificity the cell must allelically exclude the expression of one isotype and totally, biallelically, exclude the expression of the other. Such a situation is without precedent in other gene systems, and must reflect the evolutionary importance of maintaining the monospecificity of B lymphocytes. Studies of κ- and λ-producing cells and cell lines (HIETER et al. 1981; KORSMEYER et al. 1981) have shown that most κ-producing cells, although they may have undergone an abortive rearrangement of one κ allele, have maintained their λ light-chain gene loci in germ-line context. Cells producing a λ light chain, on the other hand, have usually abortively rearranged their κ light-chain gene loci and have often deleted C_κ on the abortive alleles. The deletion of an abortively rearranged C_κ gene appears to involve site-specific recombination between a "uniform deleting element" and a palindromic sequence in the J_κ–C_κ intron, and may be mediated by the same recombinational enzymes involved in V_κ–J_κ joining (DURDIK et al. 1984; SIMINOVITCH et al. 1985). These findings have led to the hypothesis that there is a hierarchy in the light-chain gene rearrangement process. A pre-B cell first attempts light-chain rearrangement on one of its κ chain alleles and if unsuccessful, repeats the attempt on the other κ allele. Having failed productively to rearrange both κ alleles, the cell would begin rearrangement of the λ locus, first on one allele, then the other. Abortively rearranged κ chain genes would suffer deletion sometime during the rearrangement cascade. Successful rearrangement at any step along the way would lead to production of a functional light chain and the expression of a complete IgM molecule on the surface of what has now become a B cell. More importantly, in terms of regulation of the rearrangement process, further light-chain gene rearrangements would cease.

A question of interest is why heavy- and light-chain genes are rearranged asynchronously. We have hypothesized (BURROWS et al. 1979) that antibody diversity might be increased by these asynchronous rearrangements. Subsequent analysis of the structure of rearranged Ig genes has made this hypothesis even more reasonable. V_H diversity is generated not only by the random association of different *V, D,* and *J* gene segments, but also by junctional and *N* region diversity, generated during the joining process, that affects the third hypervariable region (review in TONEGAWA 1983). Thus each rearranged V_H is potentially unique. By fixing a unique V_H rearrangement in a pre-B cell and then allowing its association with a broad spectrum of V_L genes in the B-cell progeny the repertoire would be increased. No *N* region diversity (e.g. MAX et al. 1979; SAKANO et al. 1979) and less junctional diversity is generated by light-chain gene rearrangement, with only two gene segments being joined, thus perhaps favoring the order heavy before light.

The reason for the apparent hierarchy of light-chain gene rearrangement, κ before λ, is not entirely obvious. In the mouse, where only two V_λ genes exist, one might argue that if the rearrangement order were λ before κ, a prohibitively large fraction of B cells would be severely restricted in their light-chain *V* region diversity. This logic, if it is that, fails in humans, where there are more V_λ genes (e.g., see ANDERSON et al. 1984) and the ratio of κ to λ B cells is ~60:40 as opposed to ~95:5 in the mouse.

3 Isotype Switching in Pre-B Cells

The immunoglobulin isotype switch is the process whereby a rearranged *VDJ* exon originally expressed with C_μ becomes expressed with another C_H isotype. The molecular mechanism for the switch (reviewed in HONJO 1983; BURROWS and COOPER 1984) has been studied extensively in plasmacytomas and hybridomas and involves a region-specific recombination between switch regions that exist 5′ to all the C_H genes except C_δ. This recombination event results in the deletion of C_μ and any other C_H genes 5′ of the newly expressed C_H gene, which then becomes the first C_H gene 3′ of the rearranged *VDJ*. Several areas of controversy, reviewed in more detail elsewhere (Burrows and Cooper 1984), still exist concerning isotype switching. These include (1) the question of whether the switch in B cells is induced by T cells or is part of the intrinsic differentiation program of some members of a B-cell clone; (2) whether switch recombination and DNA deletion are an absolute prerequisite for the expression of γ, α, or ε isotypes or whether these isotypes can be expressed while their respective C_H genes are in germ-line context; and (3) when during the differentiation of B-lineage cells isotype switching can occur. Plasma cells secreting the different isotypes have of course switched, and a small percentage of B lymphocytes bear IgG or IgA isotypes on their surfaces, often in conjunction with IgM. Thus, there is general agreement that isotype switching can occur as early as the B-cell stage of differentiation. However, there is clear discordance between the results obtained with normal and with transformed cells in terms of whether isotype switching can occur even earlier in B-lineage cells.

There have been several reports of the occurrence of more than one heavy chain isotype in A-MuLV-transformed pre-B cell lines (Burrows et al. 1981; Alt et al. 1982a; Akira et al. 1983) and in human pre-B leukemias (Vogler et al. 1981; Kubagawa et al. 1983). In the case of one of the A-MuLV lines it was shown by cellular subcloning that the cells were actually undergoing isotype switching in vitro (Burrows et al. 1983). Two aspects of the switch in the leukemic cells from both species are noteworthy. First, there is a clear isotype preference. Most switched A-MuLV-transformed cells make γ_{2b}. In human pre-B leukemias expressing more than one isotype the second isotype is usually γ_1, occasionally γ_4. Interestingly, in the human leukemias, the switch appears coupled to light-chain gene expression since the γ pre-B cells express a κ light-chain, but no surface Ig, while the μ pre-B cells are light-chain-negative. Whether there is a causal relationship between leukemogenesis and both isotype switching and light-chain expression remains to be elucidated.

Although switching in leukemic pre-B cells occurs, albeit rarely, there have been no reports of isotype switching in normal pre-B cells, even when the existence of such cells was a major focus of investigation (Kubagawa et al. 1982). The only potential exception has been the finding of small numbers of "normal" γ-producing pre-B cells in the marrow of a patient with X-linked agammaglobulinemia (Kubagawa and Cooper, unpublished observation). The issue of isotype switching in pre-B cells is not a trivial one in terms of our understanding of the normal regulation of the process, since the switch in pre-B cells, which lack an Ig receptor for antigen, would of necessity be antigen-independent. The reason for the discrepancy between results with normal and with transformed pre-B cells is not known. One could invoke several ad hoc explanations, e.g., more extensive proliferation and extended residency in the pre-B compartment by the leukemic cells or switching by normal cells at too low a frequency to be convincingly demonstrable. None of these explanations would explain either the isotype preference in the switch or the absence of switching in the majority of both human and mouse pre-B leukemias.

4 Are Some Immunodeficiency Diseases Due to Defects in Ig Gene Rearrangement?

4.1 Severe Combined Immunodeficiency

The complex series of Ig gene rearrangements that a stem cell must undergo on its way to becoming an effector cell of the B lineage, the antibody-secreting plasma cell, offers several stages where defects in the rearrangement process could lead to immunodeficiencies of varying phenotypes and severity. One of the most devastating of these diseases is termed severe combined immunodeficiency (SCID), since it affects both B- and T-lineage cells, leading to a profound deficiency in both humoral and cell-mediated immunity (see review in ROSEN et al. 1984). Human SCID is not a single disease entity, but a variety of diseases grouped together on the basis of clinical phenotype. The molecular basis for these diseases is for the most part unknown, but in ~10% of cases the defect has been traced to a deficiency in an enzyme important in nucleotide metabolism, e.g., adenosine deaminase (ADA) or polynucleotide phosphorylase (PNP) (see review in ROSEN et al. 1984). Why cells of lymphoid lineage should be so susceptible to deficiencies of these enzymes is unknown.

A mutant mouse strain has been described with some of the major characteristics of human SCID, i.e., absence of functional T and B cells (BOSMA et al. 1983; DORSHKIND et al. 1984). These mice have normal ADA and PNP levels, and the nature of the single gene defect that leads to a deficiency in both lineages is unknown. A defect in the lymphopoietic microenvironment can probably be excluded as a cause of murine SCID, since the mice can be at least partially repopulated with a source of normal stem cells (CUSTER et al. 1985). A central defect involving the transition of hemopoietic stem cells to lymphoid stem cells might explain how a single mutation could effect both lineages; however, we think this is not the case: murine SCID is a "leaky" mutation. Small numbers of thymocytes and splenic T cells are found in these mice, and older mice often develop thymic T-cell lymphomas (CUSTER et al. 1985). Analysis of an uncloned cell line derived from one such lymphoma revealed mRNA for both α and β chains of the T-cell receptor, and T_β gene rearrangements suggesting monoclonality (McNagny, Kearney, and Burrows, unpublished observations). SCID mice also have bone marrow cells, presumably of early B lineage, that can be transformed with A-MuLV (KINCADE and PHILLIPS 1985). Relatively high levels of "nonspecific" Ig can be found in the serum of some older mice. Analysis of Ig produced by hybridomas derived from leaky SCIDs revealed a striking lack of heterogeneity (SOLVASON et al. 1987). This latter observation is particularly relevant if our concept of lymphoid stem cell potential is correct: if occasional stem cells managed to "leak" through, one would expect that both B and T lineages would eventually be restored, and that Ig produced would show nearly normal heterogeneity. All of these observations argue against a defect involving lymphoid stem cell generation. Given what is known about the gene rearrangements necessary for production of Ig, and what is becoming known about the rearrangements involved in generating the T-cell antigen receptor, one could postulate a more peripheral defect resulting

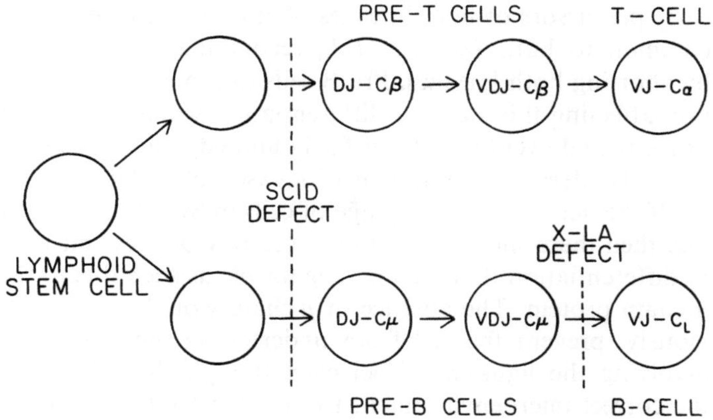

Fig. 1. Ordered gene rearrangement during B- and T-cell ontogeny and its disruption in immunodeficiency disease. Cells committed to the B lineage first undergo rearrangement of the heavy-chain locus, $D \rightarrow J$, followed by $V \rightarrow DJ$. Production of μ heavy-chain protein identifies the cell as pre-B. Expression of an immunoglobulin receptor for antigen on the surface of B cells requires rearrangement of the light (L) chain loci. During T-cell differentiation within the thymus the β chain of the T-cell receptor is expressed before the α chain. Rearrangement of β chain genes is analogous to that of Ig heavy-chain genes, being a two-step process, $D \rightarrow J$ followed by $V \rightarrow DJ$. Subsequent rearrangement of an α chain gene may be a single-step process, since no D_α gene segments have yet been identified, and it is known that V_α can recombine directly with J_α. For reasons discussed in the text, we believe that the defect in murine severe combined immunodeficiency (*SCID*) will be a common element involved in $D \rightarrow J$ joining in both B- and T-lineage cells. Some cases of human X-linked agammaglobulinemia (*X-LA*) may be due to an inability to rearrange light-chain genes or a propensity to rearrange them incorrectly

in the same SCID phenotype. The receptor for antigen on T cells is a heterodimer consisting of α and β chains (reviewed in HOOD et al. 1985). The expression of β chains requires the rearrangement of V_β, D_β, and J_β gene segments and precedes α chain expression during thymic ontogeny (Fig. 1). Both the series and the sequence of gene rearrangements required for the generation of the T-cell receptor are remarkably similar to those required for Ig gene expression. Extrapolating from what is thought to occur at the Ig loci, a defect in $D_\beta \rightarrow J_\beta$ joining could abort the development of the entire T-cell lineage. The precise mechanism for $D \rightarrow J_H$ joining is unknown, but it is thought that conserved heptamer/nonamer sequences 5′ to J_H are recognition sequences for a recombinase (reviewed in TONEGAWA 1983). Similar conserved sequences exist 5′ to J_β. The nonamer is nearly identical to that for J_H, but a portion of the heptamer is unique, and could provide T-cell specificity for a $D_\beta J_\beta$ recombinase (reviewed in DAVIS 1985). There is also evidence that the substrate specificity of this enzyme is not absolute, since some T-cell lines have undergone *DJ*, although not *VDJ* rearrangements (e.g. KUROSAWA et al. 1981), and some B-cell tumors have undergone DJ_β rearrangement (PELICCI et al. 1985 and references therein; TAWA et al. 1985). It is conceivable that the *DJ* recombinase is a multimeric enzyme. One subunit would provide specificity for either D_H or D_β and be encoded by separate genes active in either B- or T-lineage cells. Another subunit would possess endonuclease and/or ligase activity and would be expressed in early precursors of both lineages. A mutation in the gene encoding the subunit common to both DJ_β and DJ_H recombinases would cause the SCID defect by aborting both lineages. Pre-B cells that managed to "leak through" a mutation affecting this stage of differentiation would be expected to generate B-cell progeny and eventually Ig of fairly limited heterogeneity, as was observed with the SCID-derived hybridomas (SOLVASON et al. 1987).

If the defect in the B lineage of cells in SCID mice is at the level of DJ_H joining, then there should exist in the marrow of these mice cells at the pre-B stage of differentiation that are unrecognizable as such due to their failure to produce μ chain protein. The absence of μ chain would, by the rationale discussed previously, prevent the cell from undertaking the next series of rearrangements involving the light-chain loci even though the cells would be poised and in the correct microenvironment to undergo light-chain gene rearrangements. We predicted from this line of reasoning that introduction of a correctly rearranged μ heavy-chain gene into these SCID pre-B cells would bypass the SCID defect, allowing the cells to rearrange and express a light-chain gene. From a technical point of view, the simplest way to introduce a gene into normal cells, as opposed to cell lines, is by fusion. We therefore fused a μ-producing, pre-B-cell-derived hybridoma with bone marrow cells from SCID mice and examined the resulting hybrids for light-chain expression. No light chains were expressed (BURROWS, NICHOLSON, LI, and KEARNEY, unpublished observations). Unfortunately, in this type of experiment only a positive result would have been meaningful, since the lack of light-chain rearrangement could have been due to a number of factors unrelated to the original hypothesis, e.g., inhibition of light-chain gene rearrangement by the fusion partner. Creation of transgenic SCID mice microinjected with a rearranged μ chain gene should be the cleanest way to test this hypothesis.

4.2 X-Linked Agammaglobulinemia

X-linked agammaglobulinemia (X-LA) is a disorder characterized by severe depression of humoral immunity, and greatly decreased numbers of circulating B-lymphocytes. However, most of these patients have normal numbers of pre-B cells in their bone marrow, suggesting that the defect is in the pre-B/B-cell transition (PEARL et al. 1978). In one study (SCHWABER et al. 1983), hybridomas derived by fusing an unusual human hybridoma variant with cells from X-LA bone marrow were of pre-B cell phenotype (μ-chain-positive, light-chain-negative) but the μ chain protein and the μ chain mRNA were smaller than normal. Although the mRNA was not sequenced, its size and that of the protein were compatible with $DJ\text{-}C_\mu$ transcripts and translation products, as was the failure of an anti-V_H sera to stain X-LA pre-B cells. These studies would suggest that the X-LA defect is in $V \rightarrow DJ$ joining, failure of which would abort the B lineage for reasons discussed previously. Our own studies do not support this model for the pathogenesis of the disease. We have found X-LA pre-B cells that do react with monoclonal anti-V_H antibodies. In addition, X-LA pre-B cells that have been immortalized by Epstein–Barr virus (EBV) transformation have undergone heavy-chain gene rearrangement and make normal-sized μ heavy chains (Cooper et al. 1986). The reason for this discrepancy is unclear. Like SCID, X-LA is probably a heterogeneous disorder, complicating the comparison of data derived from different patients.

In EBV-transformed pre-B cell lines (μ-chain-positive, light-chain-negative) derived from three X-LA patients, the κ light-chain genes are in germ-line configuration. However, from one of our patients with X-LA, EBV-transformed pre-B cells lines with the same μ-chain-positive, light-chain-negative Ig phenotype showed abortive κ chain gene rearrangement (Kubagawa et al. 1986). Bone marrow cells from this patient were transformed on two separate occasions. In both instances, although the cultures were pauciclonal on the basis of J_H rearrangements, there was only a single κ chain gene rearrangement. The rearrangement was not the same in the two cultures. We are currently subcloning the cells to see if a common abortive κ chain gene rearrangement is associated with multiple productive J_H rearrangements. These observations suggest that in X-LA patients with pre-B cells, the defect is in the light-chain gene rearrangement process: in some patients, pre-B cells are unable to rearrange their light-chain genes, in others they do so incorrectly.

Acknowledgements. We would like to thank Max Cooper, John Kearney, Paul Kincade, Jim Downing, and Gary Borzillo for helpful discussions and Ann Brookshire for preparing the manuscript. This work was supported by grants CA 16673 and CA 13148 from the National Cancer Institute and an award from the Life and Health Insurance Medical Research Fund.

References

Akira S, Sugiyama H, Yoshida N, Kikutani H, Yamamura Y, Kishimoto T (1983) Isotype switching in murine pre-B cell lines. Cell 34:545–556

Alt FW, Rosenberg N, Lewis S, Thomas E, Baltimore D (1981) Organization and reorganization of immunoglobulin genes in AMuLV transformed cells: rearrangement of heavy but not light chain genes. Cell 27:381–390

Alt FW, Rosenberg N, Casanova RJ, Thomas E, Baltimore D (1982a) Immunoglobulin heavy chain class switching and inducible expression in an A-MuLV transformed cell line. Nature 296:325–331

Alt FW, Rosenberg N, Ena V, Siden E, Baltimore D (1982b) Multiple immunoglobulin heavy-chain gene transcripts in Abelson murine leukemia virus-transformed lymphoid cell lines. Mol Cell Biol 2:386–400

Alt FW, Yancopoulos GD, Blackwell TK, Wood C, Thomas E, Boss M, Coffman R, Rosenberg N, Tonegawa S, Baltimore D (1984) Ordered rearrangement of immunoglobulin heavy chain variable region segments. EMBO J 3:1209–1219

Anderson MLM, Szanjnert MF, Kaplan JC, McColl L, Young BD (1984) The isolation of a human Ig Vλ gene from a recombinant library of chromosome 22 and estimation of its copy number. Nucleic Acids Res 12:6647–6661

Bosma GC, Custer RP, Bosma MJ (1983) A severe combined immunodeficiency mutation in the mouse. Nature 301:527–530

Burrows PD, Cooper MD (1984) The immunoglobulin heavy chain class switch. Mol Cell Biochem 63:97–111

Burrows PD, LeJeune M, Kearney JF (1979) Evidence that murine pre-B cells synthesize μ heavy chains but no light chains. Nature 280:838–841

Burrows PD, Beck GB, Wabl MR (1981) Expression of μ and γ immunoglobulin heavy chains in different cells of a cloned mouse lymphoid line. Proc Natl Acad Sci USA 78:564–568
Burrows PD, Beck-Engeser GB, Wabl MR (1983) Immunoglobulin heavy-chain isotype switching in a pre-B cell line is accompanied by DNA rearrangement. Nature 306:243–246
Coffman RL, Weissman IL (1984) Immunoglobulin gene rearrangement during pre-B cell differentiation. J Mol Cell Immunol 1:31
Cooper MD (1981) Pre-B cells: normal and abnormal development. J Clin Immunol 1:81–89
Cooper MD, Lawton AR (1974) The development of the immune system. Sci Am 231:5–72
Cooper MD, Burrows PD, Kubagawa H (1986) Ontogeny of B cells and their abnormal development in immunodeficiency diseases. In: Aiuti F, Cooper MD, Good RA, Rosen F (eds) Recent Advances in immunodeficiencies, Vol. 28, Raven, New York, pp 19–29
Croce CM, Nowell PC (1985) Molecular basis of human B cell neoplasia. Blood 65:1–7
Custer RP, Bosma GC, Bosma MJ (1985) Severe combined immunodeficiency (SCID) in the mouse: pathology, reconstitution, neoplasms. Am J Pathol 120:464–477
Davis MM (1985) Molecular genetics of the T cell receptor β chain. Annu Rev Immunol 3:537–560
Dorshkind K, Keller GM, Phillips RA, Miller RG, Bosma GC, O'Toole M, Bosma MJ (1984) Functional status of cells from lymphoid and myeloid tissues in mice with severe combined immunodeficiency disease. J Immunol 132:1804–1808
Durdik D, Moore MW, Selsing E (1984) Novel κ light-chain gene rearrangements in mouse λ light chain-producing cells. Nature 307:749–752
Hagiya M, Davis DD, Takahashi T, Okuda K, Raschke WC, Sakano H (1986) Two types of immunoglobulin-negative Abelson murine leukemia virus-transformed cells: implications for B-lymphocyte differentiation. Proc Natl Acad Sci USA 83:145–149
Hieter PA, Korsmeyer SJ, Waldmann TA, Leder P (1981) Human immunoglobulin light chain genes are deleted or rearranged in λ-producing B cells. Nature 290:368–372
Holmberg D, Lundkvis I, Forni L, Ivars F, Coutinho A (1985) Absence of immunoglobulin heavy chain expression results in altered kappa-lambda light chain ratios. J Mol Cell Immunol 2:51–56
Honjo T (1983) Immunoglobulin genes. Annu Rev Immunol 1:499–528
Hood L, Kronenberg M, Hunkapiller T (1985) T cell antigen receptor and the immunoglobulin supergene family. Cell 40:225–229
Kemp D, Harris AW, Adams JM (1980) Transcripts of the immunoglobulin C_μ gene vary in structure and splicing during lymphoid development. Proc Natl Acad Sci USA 77:7400–7404
Kincade PW (1981) Formation of B lymphocytes in fetal and adult life. Adv Immunol 31:177–245
Kincade PW, Phillips RA (1985) B lymphocyte development. Fed Proc 44:2874–2881
Klein G (1983) Specific chromosomal translocations and the genesis of B cell derived tumors in mice and men. Cell 32:311
Korsmeyer SJ, Hieter PA, Ravetchm JV, Poplack DG, Waldmann TA, Leder P (1981) Development hierarchy of immunoglobulin gene rearrangements in human leukemic pre-B cells. Proc Natl Acad Sci USA 78:7096–7100
Kubagawa H, Gathings WE, Levitt D, Kearney JF, Cooper MD (1982) Immunoglobulin isotype expression of normal pre-B cells as determined by immunofluorescence. J Clin Immunol 2:264–269
Kubagawa H, Mayumi M, Crist W, Cooper M (1983) Immunoglobulin heavy chain switching in pre-B leukaemias. Nature 301:340–342
Kubagawa H, Burrows PD, Cooper MD (1986) Molecular and genetic aspects of human pre-B cells (Abstr). 6th International Congress of Immunology, July 7, 1986, Toronto
Kurosawa Y, von Boehmer H, Hass W, Sakano H, Traunecker A, Tonegawa S (1981) Identification of D segments of immunoglobulin heavy chains and their rearrangement in thymocytes. Nature 290:565–570
Leder P, Batey J, Lenoir G, Moulding C, Murphy W, Potter H, Stewart T, Taub R (1983) Translocations among antibody genes in human cancer. Science 222:765–771
Lennon GG, Perry RP (1985) C_μ-containing transcripts initiate heterogeneously within the IgH enhancer region and contain a novel 5′-nontranslatable exon. Nature 318:475–478
Levitt D, Cooper MD (1980) Mouse pre-B cells synthesize and secrete μ heavy chains but not light chains. Cell 19:617–625
Maki R, Kearney J, Paige C, Tonegawa S (1980) Immunoglobulin gene rearrangement in immature B cells. Science 209:1366–1369

Max E, Seidman J, Leder P (1979) Sequences of five potential recombination sites encoded close to an immunoglobulin κ constant region gene. Proc Natl Acad Sci USA 76:3450–3454

Melchers F, Andersson J, Phillips RA (1976) Ontogeny of murine B lymphocytes: development of Ig synthesis and reactivities to mitogens and to anti-Ig antibodies. Cold Spring Harbor Symp Quant Biol 41:147–158

Nelson KJ, Haimovich J, Perry R (1983) Characterization of productive and sterile transcripts from the immunoglobulin heavy-chain locus: processing of μ_m and μ_s mRNA. Mol Cell Biol 3:1317–1322

Osmond DG, Nossal GJV (1974) Differentiation of lymphocytes in mouse bone marrow. II. Kinetics of maturation and renewal of antigen-binding cells studies by double labeling. Cell Immunol 13:132–145

Owen JJT, Cooper MD, Raff MC (1974) In vitro generation of B lymphocytes in mouse fetal liver: a mammalian "bursa equivalent". Nature 249:361–363

Pearl ER, Vogler LB, Okos AJ, Crist WM, Lawton AR, Cooper MD (1978) B lymphocyte precursors in human bone marrow: an analysis of normal individuals and patients with antibody-deficiency states. J Immunol 120:1169–1175

Pelicci P, Knowles DM, Dalla-Favera R (1985) Lymphoid tumors displaying rearrangements of both immunoglobulin and T cell receptor genes. J Exp Med 162:1015–1024

Pernis B, Forni L, Amante L (1970) Immunoglobulin spots on the surface of rabbit lymphocytes. J Exp Med 132:1001–1008

Perry RP, Kelley DE, Coleclough C, Kearney J (1981) Organization and expression of immunoglobulin genes in fetal liver hybridomas. Proc Natl Acad Sci USA 78:247–251

Raff MC, Steinberg M, Taylor RB (1970) Immunoglobulin determinants on the surface of mouse lymphoid cells. Nature 225:553–554

Raff MC, Megson M, Owen JJT, Cooper MD (1976) Early production of intracellular IgM by B lymphocyte precursors in the mouse. Nature 259:224–226

Reth MG, Alt FW (1984) Novel immunoglobulin heavy chains are produced from DJH rearrangments in lymphoid cells. Nature 312:418–423

Reth MG, Ammirati P, Jackson S, Alt FW (1985) Regulated progression of a cultured pre-B cell line to the B-cell state. Nature 317:353–355

Ritchie KA, Brinster RL, Storb U (1984) Allelic exclusion and control of endogenous immunoglobulin gene rearrangement in κ transgenic mice. Nature 312:517–520

Rosen FS, Cooper MD, Wedgwood RJP (1984) The primary immunodeficiencies. N Engl J Med 311:235–242, 300–310

Rusconi S, Koehler G (1985) Transmission and expression of a specific pair of rearranged immunoglobulin μ and κ genes in a transgenic mouse line. Nature 314:330–334

Ryser JE, Vassalli P (1974) Mouse bone marrow lymphocytes and their differentiation. J Immunol 113:719–728

Sakano H, Huppi K, Heinrich G, Tonegawa S (1979) Sequences at the somatic recombination sites of immunoglobulin light-chain genes. Nature 280:288–294

Sakano H, Kurosawa Y, Weigert M, Tonegawa S (1981) Identification and nucleotide sequence of a diversity DNA segment (D) of immunoglobulin heavy chain genes. Nature 290:562–565

Schwaber J, Molgaard H, Orkin SH, Gould HJ, Rosen FS (1983) Early pre-B cells from normal and X-linked agammaglobulinemia produce $C\mu$ without an attached V_H region. Nature 304:355–358

Siden EJ, Baltimore D, Clark D, Rosenberg NE (1979) Immunoglobulin synthesis by lymphoid cells transformed in vitro by Abelson murine leukemia virus. Cell 16:389–390

Siminovitch KA, Bakhsi A, Goldman P, Korsmeyer SJ (1985) A uniform deleting element mediates the loss of κ genes in human B cells. Nature 316:260–262

Solvason N, Fried M, Bosma MJ, Kearney JF (1987) Hybridomas constructed from leaky C.B17-SCID are clonally related. (Abst.) Fed Proc

Storb U, Denis KA, Brinster RL, Witte ON (1985) Pre-B cells in κ-transgenic mice. Nature 316:356–358

Sugiyama H, Akira S, Kikutani H, Kishimoto S, Yamamura Y, Kishimoto T (1983) Functional V region formation during in vitro of a murine immature B precursor cell line. Nature 303:812–815

Tawa A, Hozumi N, Minden M, Mak TW, Gelfand EW (1985) Rearrangement of the T cell receptor β chain gene in non-T cell, non-B cell acute lymphoblastic leukemia of childhood. N Engl J Med 313:1033–1037

Tonegawa S (1983) Somatic generation of antibody diversity. Nature 302:575–581

Vogler LB, Preud'homme JL, Seligmann M, Gathings WE, Crist WM, Cooper MD, Bollum FJ (1981) Diversity of immunoglobulin expression in leukemic cells resembling B lymphocyte precursors. Nature 290:339–341

Wabl M, Burrows PD, von Gabain A, Steinberg C (1985) Hypermutation at the Ig heavy chain locus in a pre-B cell line. Proc Natl Acad Sci USA 82:479–482

Weaver D, Costantini F, Imanishi-Kari T, Baltimore D (1985) A transgenic immunoglobulin μ gene prevents rearrangement of endogenous gene. Cell 42:117–127

Yaoita Y, Matunami N, Choi CY, Sugiyama H, Kishimoto T, Honjo T (1983) The D-J_H complex is an intermediate to the complete heavy chain region gene. Nucleic Acids Res 11:7303

Cloning of Murine B-Lymphocyte Differentiation Stage-Specific Genes

N. SAKAGUCHI, C.N. BERGER, and F. MELCHERS

1 Introduction

During differentiation of eukaryotic cells with specialized functions selected genes must be expressed which encode structures for these functions. We expect that the expression of these selected genes will occur in the differentiated cell type only, and in no other cell of the body, if the corresponding structures and functions are unique for the type of differentiated cell. Examples of selected gene expression are known in blood cells. Globin is expressed in red blood cells, the T-cell receptor for antigen in T cells, and immunoglobulin (Ig) in B cells only.

Several differentiation stages of precursor B cells (pre-B cells) can be distinguished in the generation of murine B lymphocytes from stem cells. They are characterized mainly by the context in which the immunoglobulin heavy (H) and light (L) chain genes are found on the chromosomes harboring them (TONEGAWA 1983). The most immature pre-B cell appears to be one which has rearranged D_H segments onto J_H segments, followed by one in which V_H segments are rearranged to D_HJ_H segments (ALT et al. 1984; SUGIYAMA et al. 1983; YANCOPOULOS et al. 1984). Next, pre-B cells appear in which not only $V_HD_HJ_H$ has been rearranged, but also V_L to J_L. It is generally accepted that κ genes rearrange prior to λ genes (HIETER et al. 1981). At the end of this process Ig μH-chain- and L-chain-expressing B cells appear that are resting and will respond to antigens and mitogens.

Immunoglobulins on B cells, as well as antigen-specific T-cell receptors on T cells, are known to exhibit specific antigen recognition which results in prolif-

Basel Institute for Immunology, Grenzacherstrasse 487, 4058 Basel, Switzerland

Current Topics in Microbiology and Immunology, Vol. 135

eration and/or the development of effector functions of these lymphocytes. It is also clear that other molecules must be expressed for a complete response of lymphocytes to antigen. Some of these molecules could be selectively expressed, as T-cell receptors and Ig molecules are in T and B cells respectively. The isolation of genes specific for given stages of the B-cell lineage could, therefore, lead to identification of molecules which cooperate in the specialized, differentiated function of these cells within the immune system.

Here, we review some of the approaches which have been taken to characterize B-lineage- and differentiation-stage-specific molecules and to isolate selectively expressed genes. We describe the ways in which such genes can be obtained through differential hybridization of cDNA libraries obtained from cells at differeant stages and of different lineages of differentiation.

2 B-Lymphoid Lineage-Specific Molecules and Functions

Specific structures on pre-B, B, or plasma cells can be distinguished by virtue of the specific binding of ligands such as antigen or growth and maturation factors (Table 1). Many of them have been characterized as protein molecules. For some the corresponding genes have also been cloned and sequenced. Binding of the ligands or antibodies to some of these structures influences the functions of B cells which, in all cases so far studied, have been found to be connected with the control of proliferation and maturation of the cells.

Most of the structures listed in Table 1 are not expressed throughout development from the earliest pre-B cell to a plasma cell. Structures recognized by the monoclonal antibodies J11d, AA4, and GF-1 are predominantly expressed on pre-B, but little or not at all on B and plasma cells (Bruce et al. 1981; McKearn et al. 1984; McKearn and Rosenberg 1985). Lyt-1 and GF-1 are expressed only on subsets of pre-B cells and on some B cells. Lyt-1 marks pre-B cells with autoimmune potential (Hayakawa et al. 1985). On the other hand, Lyb-2 (Sato and Boyse 1976; Shen et al. 1981; Subbarao and Mosier 1983), Lyb-3 (Huber et al. 1977; Cone et al. 1978), Lyb-5, and Lyb-7, NiMR-2 and NiMR-3 (Parkhouse 1985), and reactivity for LPS are mainly expressed on mature B cells. PC-1 and PC-2, finally, are found only on Ig-secreting plasma cells, which tend to lose many of the mature B-cell-specific markers as they mature to plasmablasts and, eventually, to plasma cells. The widest distribution from pre-B cells to plasmablasts (Osmond 1979; Sitia et al. 1979) is found for Ig H chain expression, for MHC class II antigens, and for Fc receptors. The analysis of the expression of these surface markers has been facilitated by the use of a series of pre-B lymphomas, B lymphomas, and plasmacytomas. In general, a remarkable concordance of Ig expression phenotype, morphological character, functional state, and surface marker phenotype has been observed.

Many of the structures listed in Table 1 are expressed not only on cells of the B lineage, but also on other cells of the body. They are, therefore, unlikely to account for any functions unique to B-lymphocytes. The limited amount

Table 1. Murine B cell lineage-related structures and functions

Structure	Function	Molecule(s) identified	Gene(s) cloned and sequenced	Expression as protein in other than B-lineage cells[a]	References
IgH+L	Antigen binding in virgin and memory B cells, growth regulation	+	+	No	Möller (1961); Pernis et al. (1965); Kincade et al. (1970); Warner (1974); Kishimoto and Ishizakaj (1975); Parker et al. (1979); De Franco et al. (1982); Melchers and Andersson (1984)
Fc receptor	Binding of secreted Ig (often as complexes with antigen), growth regulation	+60 KD +47 KD	+ +	Probably yes	Unkeless et al. (1981)
Complement CRI receptor	C3b binding	+160 KD +190 KD +220 KD +250 KD	+	Yes	Ross and Medof (1985)
Complement CR2 receptor	C3d binding growth regulation	+140 KD	+	Yes	Lambris et al. (1981) Melchers et al. (1985)
MHC class II (Ia antigen)	T-B interaction, antigen presentation, growth regulation	+	+	Yes	Katz et al. (1973) Sprent (1978) Lemke et al. (1981)
LPS receptor	LPS binding, growth regulation	–	–	Yes	Forsgren et al. (1984); Andersson et al. (1972); Coutinho and Möller (1975); Wetzel and Kettman (1981); Melchers and Andersson (1984)
Lyb-2	Growth regulation	+40 to 50 KD	–	No	Sato and Boyse (1976) Shen et al. (1981) Subbarao and Mosier (1983)
Lyb-3	Growth regulation	–	–	No	Huber et al. (1977) Cone et al. (1978)
Lyb-5	?	–	–	No	Ahmed et al. (1977) Subbarao and Mosier (1983)
Lyb-7	?	–	–	No	Subbarao and Mosier (1983)

[a] "No" indicates that expression in other cells has not yet been observed

Table 1 (continued)

Structure	Function	Molecule(s) identified	Gene(s) cloned and sequenced	Expression as protein in other than B-lineage cells[a]	References
B220 (14.8, Ly 5)	?	+220 KD	+	Yes	COFFMAN and WEISSMAN (1981 a) COFFMAN and WEISSMAN (1981 b) KINCADE et al. (1981) THOMAS et al. (1985) SHEN et al. (1985)
J11d	?			Yes	BRUCE et al. (1981)
AA-4 GF-1	? ?			Yes	MCKEARN et al. (1984) MCKEARN and ROSENBERG (1985)
BP-1	?	+	–	No	COOPER et al. (1986)
PC-1	?	+230 KD	+	Yes	TAKAHASHI et al. (1970) STEARNE et al. (1985) DRIEL (1985)
PC-2	?			No	TADA et al. (1980)
Lyt-1	?	+	+	Yes	HAYAKAWA et al. (1985)
NiMR-2	Growth regulation	–	–	No	PARKHOUSE (1985)
NiMR-3	Growth regulation	–	–	No	PARKHOUSE (1985)
BLA-1	?	+53 KD	–	No	HARDY et al. (1984)
BLA-2	?		–	No	HARDY et al. (1984)
myc	Growth regulation and	+	+	Yes	MARCU et al. (1983)
myb	deregulation	+	+	Yes	ZIMMERMAN et al. (1986)
bcl-1		–	+	Yes	MOUNTZ et al. (1984)
bcl-2		–	+	Yes	TSUJIMOTO et al. (1985a, b)

of information about and characterization of B-lineage-specific markers and their functions listed in Table 1 demonstrates the need for additional approaches to isolate and characterize B-lineage-specific molecules. One such approach is the construction of cDNA libraries from differentiated cell types, followed by subtractive hybridization. The efficacy of this method was proven by the isolation of the genes for the α, β, and γ chains of the T-cell receptor (HEDRICK et al. 1984a, b; SAITO et al. 1984a, b; CHIEN et al. 1984) and for the genes related to X-linked immunodeficiency disorders in CBA/N mice (COHEN et al. 1985).

3 The Construction of Subtractive cDNA Libraries

The search for B-lineage-specific mRNA molecules is likely to extend the spectrum of specific structures which are expressed during B-cell differentiation. This method will detect gene products to which antibodies cannot be raised, either due to lack of immunogenicity or because they are internally sequestered.

The strategies to obtain lineage- and differentiation-stage-specific cDNA clones have been laid out by DAVIS et al. (1984). We will describe in brief the method for the construction of a murine pre-B-cell-specific cDNA library which we have undertaken.

Membrane-bound polysomal mRNA was extracted from the pre-B cell line 70Z/3 (PAIGE et al. 1978; SAKAGUCHI et al. 1980). A cDNA library, up to 3.5 kb in length, was made from this mRNA preparation. The library was then subjected to subtractive hybridization. This was done twice with an excess amount of RNA prepared from the T-cell hybridoma K62. This T-cell hybridoma was obtained by fusion of an alloreactive, H-2^k-specific helper-T-cell clone with the thymic lymphoma BW5147 (A. Rolink, unpublished experiments). The cDNA library, now enriched for B-cell-specific sequences, was made double-stranded and was then ligated into the λgt11 phage vector (YOUNG and DAVIS 1983). From selected cDNA clones, single-stranded probes were made in phage M13. These single-stranded probes could then be labeled with tritiated nucleotides and used in the situ hybridizations with single cells or tissue sections as described (BERGER 1986a, b). This in situ hybridization technique will detect as little as 50 mRNA molecules per cell. This is probably equivalent to the detection of 10^4–10^5 surface-bound or intracytoplasmic protein molecules by specific antibodies, which are expected to be made by 50–500 mRNA molecules per cell.

The pre-B cell cDNA library was initially screened for clones containing sequences expressed in pre-B cells but not in T cells, using radiolabeled cDNA preparations of 70Z/3 and of K62 cells. Clones positive on the former but negative on the latter were further screened on a set of mRNA preparations from cell lines and tumors of different stages of B-cell differentiation. The cDNA of the clone was radiolabeled and hybridized to the mRNA preparation blotted onto nitrocellulose filters. The pre-B cell, B cell, myeloma, T cell, and monocyte lines and tumors used in these studies are shown in Fig. 1.

4 The Identification of a Gene Which Is Preferentially Expressed at Pre-B Cell Stages of B-Cell Development

One clone of interest, pZ183, was originally selected because it was expressed in 70Z/3 but not in K62 cells. Figure 1 shows the dot-blot analysis done with the radioactively labeled pZ183 cDNA clone with mRNA preparations of a series of pre-B, B, plasma cell, T and monocyte cell lines and tumors. It is evident from this analysis that pZ183 is preferentially expressed in pre-B, but not in B, plasma cell, T, and monocyte lineage cells. Since the pre-B cell lines

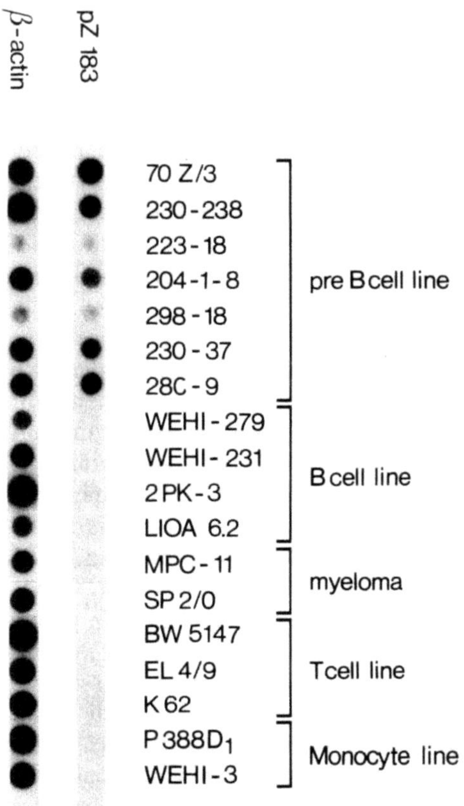

Fig. 1. Expression of pZ183 transcripts in the cell lines of different stages and lineages. Cytoplasmic RNA from various cell lines was dotted onto nitrocellulose filter for hybridization. Filters were hybridized with radioactive probe of pZ183 or β-actin and washed finally with 0.2 × SSC, 1% SDS at 65°C and exposed to film. Mature B-cell line 2PK-3, T-cell line EL4/9, and Abelson-virus-transformed pre-B cell lines, including μ gene incompletely rearranged, (D–J_H rearranged) 298 – 18 and the completely V_H–D–J_H rearranged cells 230–238, 223 – 18, 204 – 1 – 8, 230–37, and 28C – 9, were kindly provided by Drs. J. McKearn and N. Rosenberg. The lines 230–238, 233 – 18, 230–37 (C57L origin) and 298–18, 204 – 1 – 8 (BALB/c origin) were transformants from adult bone marrow. 28C – 9 was the transformant from fetal liver at day 17–19 of BALB/c origin. The B-cell lines WEH1–279, WEH1–231, L10A6.2, MPC – 11, and SP2/0, the T-cell line BW 5147, and the monocytic cell lines P388D1 and WEH1–3, were obtained as described by Corbel and Melchers (1984). The T-cell hybridoma K62 was provided by A. Rolink in our laboratory

are obtained from different mouse strains, the pZ183 sequences are not haplotype restricted. Sequences of pZ182 are expressed in a wide spectrum of pre-B cell lines, including those which have their H chain genes only incompletely rearranged ($D_H J_H$) as well as those which have H and L chains rearranged (but L not yet expressed; 70Z/3). Furthermore, expression is not restricted to Abelson-virus-transformed cell lines, since the donor cell 70Z/3 is a chemically induced lymphoma which does not contain Abelson virus.

Northern blot analysis shows that pZ183 cDNA hybridizes to mRNA of 1.2 kb (Sakaguchi et al. 1986). This size of mRNA is the same in all pre-B cell lines tested. It is, therefore, smaller than the transcript for terminal transferase (2.1 kb; LANDAU et al. 1984), a marker of precursor T and B cells, and smaller than transcripts of c-*abl* and v-*abl* (5.3 and 6.5 kb; BEN-NERIAH et al. 1986). The partial sequence of pZ183 obtained so far differs from those of Ig, MHC, c-*myc*, v-*abl*, c-*abl*, v-*fos*, c-*mos*, v-*raf*, c-Ki-*ras*, v-*erb*, and c-Ha-*ras*. The nature of the gene and its function(s) remain to be elucidated.

Detection of pZ183 transcripts in normal lymphoid cells proved to be more difficult, since by Northern dot-blot analysis a panel of mRNA preparations of adult spleen, liver, kidney, thymus, and bone marrow was found to be negative, i.e., not significantly labeled above the background hybridization. Normal pre-B cells, however, are not very frequent even in the organs in which they can be expected to be present. We therefore employed the in situ hybridization technique with tissue sections and single-cell suspensions to probe for pZ183 expression in single cells (BERGER 1986a, b). We found that only fetal liver and bone marrow contained pZ183-positive cells: 2%–4% in fetal liver of day 15–16 of gestation and 2% in adult bone marrow. We suspect on the basis of their frequency and tissue distribution, but have not yet proved, that these positive cells are normal pre-B cells.

5 Future Perspectives

Among the cDNA library obtained from 70Z/3 pre-B cells by subtractive hybridization, we have already detected many more genes which are expressed in B- but not T-lineage cells. They are likely to widen the potentialities for following the development of the B-lymphocyte lineage in vitro and in vivo. It can be expected that we will be able to elucidate the complete structure of the cDNAs and their genomic counterparts. It would be possible to express these genes and, thus, to identify the functions of these B-lineage-related or -specific gene products.

Acknowledgements. The able technical assistance of Heidi Bächtold, Denise Richterich, and Annik Peter is gratefully acknowledged. We thank Drs. Naomi Rosenberg and John McKearn for providing us with the Abelson-virus-transformed immature B-cell lines and Dr. M. Buckingham for providing us with β-actin DNA. The Basel Institute for Immunology was founded and is supported by Hoffmann-La Roche, Ltd, Basel, Switzerland.

Note Added in Proof

In the meantime, the complete nucleotide sequence of the coding region of pZ183 has been determined (Sakaguchi, N. and Melchers, F. [1986] Nature 324:579–582). Furthermore, the genomic structure of pZ183 has been elucidated (Kudo, A., Sakaguchi, N., and Melchers, F. [1987] EMBO J 6:103–107). The gene shows strong homologies to the J- and C-segments of λ-L chains, has a 5′ sequence with no apparent strong homologies to any other known sequence and is *not* rearranged during pre B and B cell development.

References

Ahmed A, Scher I, Sharrow SO, Smith AH, Paul WE, Sachs DH, Sell KW (1977) B-lymphocyte heterogeneity: development and characterization of an allo antiserum which distinguishes B-lymphocyte differentiation alloantigens. J Exp Med 145:101–110

Alt FW, Yancopoulos GD, Blackwell TK, Wood C, Thomas E, Boss M, Coffman R, Rosenberg N, Tonegawa S, Baltimore D (1984) Ordered rearrangement of immunoglobulin heavy chain variable region segments. EMBO J 3:1209–1219

Andersson J, Sjöberg O, Möller G (1972) Induction of immunoglobulin and antibody synthesis in vitro by lipopolysaccharides. Eur J Immunol 2:349–353

Ben-Neriah Y, Bernards A, Paskind M, Daley GQ, Baltimore D (1986) Alternative 5′ exons in c-*abl* mRNA. Cell 44:577–586

Berger CN (1986a) In situ hybridization of immunoglobulin specific RNA in single cells of the B lymphocyte lineage with radiolabeled DNA probes. EMBO J 5:85–93

Berger CN (1986b) In situ Hybridisierung von mRNA Molekülen in einzelnen lymphoiden Zellen und Anwendungen zur Untersuchung der Repertoires Antikörper produzierender Lymphozyten. Dissertation, University of Basle

Bruce J, Symington FW, McKearn TJ, Sprent J (1981) A monoclonal antibody discriminating between subsets of T and B cells. J Immunol 127:2496–2501

Chien Y, Becker DM, Lindsten T, Okamura M, Cohen DI, Davis MM (1984) A third type of murine T-receptor gene. Nature 312:31–35

Coffman RL, Weissman IL (1981a) A monoclonal antibody that recognizes B cells and B cell precursors in mice. J Exp Med 153:269–279

Coffman RL, Weissman IL (1981b) B220: a B cell-specific member of the T200 glycoprotein family. Nature 289:681–683

Cohen DI, Steinberg AD, Paul WE, Davis MM (1985) Expression of an X-linked gene family (XLR) in late-stage B cells and its alteration by the xid mutation. Nature 314:372–373

Cone RE, Huber B, Cantor H, Gershon RK (1978) Molecular identification of a surface structure on B cells (Lyb-3) and its relationship to B cell triggering. J Immunol 120:1733–1740

Cooper MD, Mulvaney D, Coutinho A, Canave P-A (1986) Identification of a novel cell surface molecular on early B-lineage cells. Nature, to be published

Corbel C, Melchers F (1984) The synergism of accessory cells and of soluble α-factors derived from them in the activation of B cells to proliferation. Immunol. Rev. 78:51–74

Coutinho A, Möller G (1975) Thymus-independent B-cell induction and paralysis. Adv Immunol 21:113–236

Davis MM, Cohen DI, Nielsen EA, Steinmetz M, Paul WE, Hood L (1984) Cell-type-specific cDNA probes and the murine I region: the localization and orientation of A^d_α. Proc Natl Acad Sci USA 81:2194–2198

De Franco AL, Ashwell JD, Schwartz RH, Paul WE (1982) Frequency of B lymphocytes responsive to anti-immunoglobulin. J Exp Med 155:1523–1536

Driel IRV (1985) Murine plasma cell membrane antigen PC-1: Molecular cloning of cDNA and analysis of expression. Proc Natl Acad Sci USA 82:8619–8623

Forsgren S, Pobor G, Coutinho A, Pierres M (1984) The role of I-A/E molecules in B lymphocyte activation. I. Inhibition of lipopolysaccharide-induced responses by monoclonal antibodies. J Immunol 133:2104–2110

Hardy RR, Hayakawa K, Parks DR, Herzenberg LA, Herzenberg LA (1984) Murine B cell differentiation lineages. J Exp Med 159:1169–1188

Hayakawa K, Hardy RR, Herzenberg LA, Herzenberg LA (1985) Progenitors for Ly-1 B cells are distinct from progenitors for other B cells. J Exp Med 161:1554–1568

Hedrick SM, David I, Cohen D, Nielsen EA, Davis MM (1984a) Isolation of cDNA clones encoding T cell-specific membrane associated proteins. Nature 308:149–153

Hedrick SM, Nielsen EA, Kavaler J, Cohen DI, Davis MM (1984b) Sequence relationships between putative T-cell receptor polypeptides and immunoglobulins. Nature 308:153–158

Hieter PA, Stanely J, Korsmeyer, Waldmann TA, Leder P (1981) Human immunoglobulin κ light chain genes are deleted or rearranged in λ-producing B cells. Nature 290:368–372

Huber BT, Gershon KK, Cantor H (1977) Identification of a B-cell surface structure involved in

antigen-dependent triggering: absence of this structure on B cells from CBA/N mutant mice. J Exp Med 145:10–16
Katz DH, Hamaoka T, Dorf ME, Benacerraf B (1973) Cell interaction between histocompatible T and B lymphocytes. III. Demonstration that the H-2 gene complex determines successful physiologic lymphocyte interactions. Proc Natl Acad Sci USA 70:2624–2628
Kincade PW, Lawton AR, Bockman DE, Cooper MD (1970) Suppression of immunoglobulin G synthesis as a result of antibody-mediated suppression of immunoglobulin M synthesis in chickens. Proc Natl Acad Sci USA 67:1918–1925
Kincade PW, Lee G, Paige CJ, Scheid MP (1981) Antigens displayed on murine B lymphocyte precursors. J Immunol 127:2262–2266
Kishimoto T, Ishizaka K (1975) Regulation of antibody response in vitro. IX. Induction of secondary anti-hapten IgG antibody response by anti-immunoglobulin and enhancing soluble factor. J Immunol 114:585–591
Lambris JD, Dobson NJ, Ross GD (1981) Isolation of lymphocyte membrane complement receptor type two (the C3d receptor) and preparation of receptor-specific antibody. Proc Natl Acad Sci USA 78:1828–1832
Landau NR, St John TP, Weissman IL, Wolf SC, Silverstone AE, Baltimore D (1984) Cloning of terminal transferase cDNA by antibody screening. Proc Natl Acad Sci USA 81:5836–5840
Ledbetter JA, Herzenberg LA (1979) Xenogeneic monoclonal antibodies to mouse lymphoid differentiation antigens. Immunol Rev 47:362–384
Lemke H, Hämmerling GJ, Hämmerling U (1981) Fine specificity analysis with monoclonal antibodies of antigens controlled by the major histocompatibility complex and by the Q2/TL region in mice. Immunol Rev 47:175–206
Marcu KB, Harris LJ, Stanton LW, Erikson J, Watt R, Croce CM (1983) Transcriptionally active c-*myc* oncogene is contained within NIARD, a DNA sequence associated with chromosome translocations in B cell neoplasia. Proc Natl Acad Sci USA 80:519–523
McKearn JP, Rosenberg N (1985) Mapping cell surface antigens on mouse pre-B cell lines. Eur J Immunol 15:295–298
McKearn JP, Baum C, Davie JM (1984) Cell surface antigens expressed by subsets of pre-B cells and B cells. J Immunol 132:332–339
Melchers F, Andersson J (1984) B cell activation: three steps and their variations. Cell 37:715–730
Melchers F, Melchers J, Philips RA (1976) Ontogeny of murine B lymphocytes: development of Ig synthesis and reactivities to mitogens and anti-Ig antibodies. In Cold Spring Harbor Symp Quant Biol 41:147–158
Melchers F, Erdei A, Schulz T (1985) Growth control of activated, synchronized murine B cells by the C3d fragment of human complement. Nature 317:264–267
Möller G (1961) Demonstration of mouse iso-antigens at the cellular level by the fluorescent antibody technique. J Exp Med 114:415–434
Mountz JD, Steinberg AD, Klinman DM, Smith HR, Mushinski JF (1984) Autoimmunity and increased c-*myb* transcription. Science 226:1087–1089
Osmond DG (1979) Generation of B lymphocytes in the bone marrow. In: Cooper M, et al. (eds) B lymphocyte in the immune response. Elsevier North Holland, Amsterdam, pp 63–70
Paige CJ, Kincade PW, Ralph P (1978) Murine B cell leukemia line with inducible surface immunoglobulin expression. J Immunol 121:641–647
Parker DC, Frothergill JJ, Wadsworth DC (1979) B lymphocyte activation by insoluble anti-immunoglobulin: induction of immunoglobulin secretion by a T cell-dependent soluble factor. J Immunol 123:931–941
Parkhouse M (1985) in Mechanisms of B cell neoplasia 1985. Workshop at the Basel Institute for Immunology, pp 48–54, Editiones Roche, Basel
Pernis B, Chiappino G, Kelus AS, Gell PGH (1965) Cellular localization of immunoglobulins with different allotypic specificities in rabbit lymphoid tissues. J Exp Med 122:853–876
Ross GD, Medof ME (1985) Membrane complement receptors specific for bound fragments of C3. Adv Immunol 37:217–267
Sakaguchi N, Kishimoto T, Kikutani H, Watanabe T, Yoshida N, Shimizu A, Yamawaki-Kataoka Y, Honjo T, Yamamura Y (1980) Induction and regulation of immunoglobulin expression in a murine pre-B cell line, 70Z/3. I. Cell cycle associated induction of sIgM expression and κ-chain synthesis in 70Z/3 cells by LPS stimulation. J Immunol 125:2654–2659

Sakaguchi N, Berger CN, Melchers F (1986) Isolation of a cDNA copy of an RNA species expressed in murine pre-B cells. EMBO J 5:2139–2147

Saito H, Kranz DM, Takagaki Y, Hayday AC, Eisen HN and Tonegawa S (1984a) Complete primary structure of a heterodimeric T-cell receptor deduced from cDNA sequences. Nature 309:757–762

Saito H, Kranz DM, Takagaki Y, Hayday AC, Eisen HN, Tonegawa S (1984b) A third rearranged and expressed gene in a clone of cytotoxic T lymphocytes. Nature 321:36–40

Sato H, Boyse EA (1976) A new alloantigen expressed selectively on B cells: the Lyb-2 system. Immunogenetics 3:565–572

Shen FW, Yakura H, Tung J-S (1981) Some compartments of B cell differentiation. Immunol Rev 69:69–80

Shen FW, Saga Y, Litman G, Freeman G, Tung J-S, Cantor H, Boyse EA (1985) Cloning of Ly-5 cDNA. Proc Natl Acad Sci USA 82:7360–7363

Sitia R, Abbott J, Hämmerling U (1979) The ontogeny of B lymphocytes. V. Lipopolysaccharide-induced changes of IgD expression on murine B lymphocytes. Eur J Immunol 9:859–864

Sprent J (1978) Role of H-2 gene products in the function of T helper cells from normal and chimeric mice measured in vivo. Immunol Rev 42:108–137

Stearne PA, van Driel IR, Greco B, Simpson RJ, Goding JW (1985) The murine plasma cell antigen PC-1: Purification and partial amino acid sequence. J Immunol 134:443–448

Subbarao B, Mosier DE (1983) Lyb antigens and their role in B lymphocyte activation. Immunol Rev 69:81–98

Sugiyama H, Akira S, Kikutani H, Kishimoto S, Yamamura Y, Kishimoto T (1983). Functional V region formation during in vitro culture of a murine immature B precursor cell line. Nature 303:812–815

Tada N, Kimura S, Hoffmann M, Hämmerling U (1980) A new surface antigen (PC.2) expressed exclusively on plasma cells. Immunogenetics 11:351–359

Takahashi T, Old, LJ, Boyse EA (1970) Surface alloantigens of plasma cells. J Exp Med 131:1325–1332

Thomas ML, Barclay AN, Gragnon J, Williams AF (1985) Evidence from cDNA clones that the rat leukocyte-common antigen (T200) spans the lipid bilayer and contains a cytoplasmic domain of 80000 Mr. Cell 41:83–93

Tidmarsh GT, Dailey MO, Whitlock CA, Pillemer E, Weissman IL (1985) Transformed lymphocytes from Abelson-diseased mice express levels of a B lineage transformation-associated antigen elevated from that found on normal lymphocytes. J Exp Med 162:1421–1434

Tonegawa S (1983) Somatic generation of antibody diversity. Nature 302:575–581

Tsujimoto Y, Jaffe E, Cossman J, Gorham J, Nowell PC, Croce CM (1985a) Clustering of breakpoints on chromosome 11 in human B-cell neoplasms with the t(11:14) chromosome translocation. Nature 315:340–343

Tsujimoto Y, Cossman J, Jaffe E, Croce CM (1985b) Involvement of the bcl-2 gene in human follicular lymphoma. Science 228:1440–1443

Unkeless JC, Howard F, Mellman IS (1981) Structural aspects and heterogeneity of immunoglobulin Fc receptors. Adv Immunol 31:247–270

Velardi A, Cooper MD (1984) An immunofluorescence analysis of the ontogeny of myeloid, T, and B lineage cells in mouse hemopoietic tissues. J Immunol 133:672–677

Warner NL (1974) Membrane immunoglobulins and antigen receptor on B and T lymphocytes. Adv Immunol 19:67–216

Warner NL, Daley MJ, Richey J, Spellman C (1979) Flow cytometry analysis of murine B lymphoma differentiation. Immunol Rev 48:197–216

Wetzel GD, Kettman JR (1981) Activation of murine B lymphocytes. III. Stimulation of B lymphocyte clonal growth with lipopolysaccharide and dextran sulfate. J Immunol 126:723–728

Yancopoulos GD, Desiderio SV, Paskind M, Kearney JF, Baltimore D, Alt FW (1984) Preferential utilization of the most J_H-proximal V_H gene segments in pre-B cells lines. Nature 290:368–372

Young RA, Davis RW (1983) Efficient isolation of genes by using antibody probes. Proc Natl Acad Sci USA 84:1194–1198

Zimmerman KA, Yancopoulos GD, Collum RG, Smith RK, Kohl NE, Denis KA, Nau MM, et al. (1986) Differential expression of *myc* family genes during murine development. Nature 314:780–783

Subject Index

MIX
Papier aus verantwortungsvollen Quellen
Paper from responsible sources
FSC® C105338

If you have any concerns about our products,
you can contact us on
ProductSafety@springernature.com

In case Publisher is established outside the EU,
the EU authorized representative is:
Springer Nature Customer Service Center GmbH
Europaplatz 3, 69115 Heidelberg, Germany

Printed by Libri Plureos GmbH
in Hamburg, Germany